ENGINEERING PHYSICS

Georgies Alene Asres

TAP

Toronto Academic Press

ENGINEERING PHYSICS

Georgies Alene Asres

Toronto Academic Press
4164 Lakeshore Road
Burlington ON L7L 1A4
Canada
www.tap-books.com

Email: orders@arclereducation.com

ISBN: 978-1-77956-304-0

ABOUT THE AUTHOR

Georgies Alene Asres is a dedicated materials engineer with a diverse background in materials science and engineering, electrical engineering, physics, and research. He currently serves as an associate professor of Materials Engineering at Addis Ababa University in Ethiopia, where he is also the head of Center for Materials Engineering. Georgies earned his PhD in Electrical Engineering (Electronics Materials and Devices) from the University of Oulu, Finland. He holds dual master's degrees in Functionalized Advanced Materials and Engineering from Grenoble Institute of Technology (France) and Augsburg University (Germany). With over eight years of teaching and research experience, including his tenure at Addis Ababa University, Georgies has established himself as a respected academic and researcher. In his role as an associate professor, Georgies teaches undergraduate and graduate courses in materials science and engineering. His research focuses on developing new materials for energy applications, among other topics in materials engineering. He supervises MSc and PhD students, providing guidance and feedback on their research projects. Georgies is actively involved in collaborative research with faculty members and external partners, contributing to interdisciplinary projects and grant proposals. He enriches the academic community by publishing his research findings in peer-reviewed journals and conferences, reviewing papers for scholarly publications, and serving on academic committees.

Table of Contents

List of Figures

List of Tables

PREFACE

In Engineering Physics, precision and innovation are paramount. This interdisciplinary field not only explores the fundamental laws governing the physical world but also applies these principles to create cutting-edge solutions that propel technological advancements. From the nanoscale intricacies of semiconductor devices to the macroscopic dynamics of structural engineering, engineering physics plays a crucial role in shaping modern civilization. By understanding and harnessing the principles of mechanics, optics, and materials science, engineers in this field develop tools and technologies that enhance medical diagnostics, optimize industrial processes, and revolutionize transportation systems, ensuring sustainable and efficient solutions for the challenges of tomorrow.

This book comprises eight comprehensive chapters. Beginning with Chapter 1, readers will explore foundational concepts such as relativity, quantum mechanics, and wave-particle duality, setting the stage for a deeper exploration of modern scientific thought.

Chapter 2 examines the intriguing world of quantum phenomena, from atomic structure to quantum entanglement, elucidating how these principles underpin advanced technologies like quantum computing and nanotechnology. In Chapter 3, the focus shifts to materials science, examining the electrical properties of conductors and semiconductors that are critical to developing the electronic devices driving modern technology.

The principles of electromagnetism and electrodynamics are discussed in Chapter 4, providing a thorough understanding of electromagnetic waves, Maxwell's equations, and their applications in telecommunications and electrical engineering. Chapter 5 explores the structure and behavior of crystalline solids, laying the groundwork for innovations in materials engineering and semiconductor technology.

In Chapter 6, readers delve into the properties of light and its manipulation through optical devices, essential for advancements in imaging, telecommunications, and laser technology. Chapter 7 addresses the principles governing energy transfer and the behavior of systems at both the macroscopic and microscopic scales, fundamental to fields ranging from mechanical engineering to renewable energy technologies.

Chapter 8 concludes the book by exploring the structure of atomic nuclei, radioactive decay, and nuclear reactions, highlighting their applications in energy generation and medical diagnostics.

This book is tailored for students, researchers, and professionals seeking a comprehensive exploration of engineering physics principles. By bridging theoretical knowledge with practical applications across these diverse chapters, we aim to provide a valuable resource that fosters understanding and innovation in fields as varied as materials science, electrical engineering, and quantum technologies.

—Author

CHAPTER

1 Modern Physics

LEARNING OBJECTIVES

After studying this chapter, you will be able to:

1. The failures of classical mechanics and origin of Quantum Mechanical Theory;
2. The application of quantum idea to explain black body radiation spectrum, photoelectric effect, and Compton effect;
3. The de-Broglie hypothesis (matter waves) and experimental evidence given by Davison and Germer;
4. The relation between the group velocity, phase velocity, particle velocity and velocity of light;
5. Characteristics of matter waves.

KEY TERMS FROM THIS CHAPTER

Blackbody radiation
Compton shift
de Broglie hypothesis
Group velocity
Particle velocity
Photoelectric effect
Blackbody radiation spectrum
Davisson-Germer experiment
de Broglie wavelength
Modern physics
Phase velocity
Speed of light

UNIT INTRODUCTION

Science derives from the Latin term "scientia," signifying the pursuit of knowledge. Science is the result of careful investigations and observations. Scientific methods encompass crucial elements, including modeling, logical reasoning, theoretical prediction, and systematic observations. Science is classified into two main disciplines: natural science and philosophical science. Within the realm of natural science are numerous disciplines, and physics is a prominent field among them (Gil and Solbes, 1993).

The term "physics" can be traced back to its Greek root, "physis," which means "nature." Physics, as a fundamental scientific discipline, explores the intricacies of nature and its various phenomena. It delves into the properties of matter, energy, and their intricate interactions. Having a strong foundation in science begins with a comprehensive understanding of physics. Physics is a discipline that heavily relies on careful observation and rigorous experimentation. Observations and measurements of physical phenomena serve as the ultimate benchmark for evaluating the validity of any physical theory (Chandler, 1987).

Technology uses the principles of physics to achieve practical objectives. The discovery of the steam engine revolutionized human society. Until 1933, Rutherford held the belief that extracting energy from atoms was not possible. In 1938, Hahn and Meitner made a significant breakthrough in their research on the neutron-induced fission reaction of uranium. There is a pressing need to explore alternative energy sources due to the rapid depletion of fossil fuels and the increasing demand for affordable options. Generating electricity from solar and geothermal energy sources has become a tangible achievement in recent times. However, there is still significant progress to be made in this field. Another illustration of the impact of physics on technology includes the integrated chip, commonly known as an IC. In the past three decades, the computer industry has seen major developments thanks to the advancements in integrated circuits and faster processors. Due to recent developments in manufacturing methods and cost reduction, computers have become increasingly accessible to a broader audience (Kuhn, 1961).

The ultimate objective of modern technology is to enhance the well-being of humankind. Our society is increasingly adopting a scientific mindset. Enhancing our knowledge of the basic concepts of physics will enable us to make more significant contributions to the scientific community and society at large (Giancoli, 2008).

1.1. BLACKBODY RADIATION SPECTRUM

The fundamental principles of modern physics, specifically focusing on the spectrum of blackbody radiation, the distribution of energy within this spectrum, and the governing laws that dictate blackbody radiation, are explored in this text (Boyer, 2012).

1.1.1. Origin of Modern Physics

In the late 19th century, the field of physics was dominated by Newton's laws of motion, the laws of momentum and energy conservation, the laws of thermodynamics, universal gravitation, and Maxwell's equations for magnetism and electricity. Many people believe that there are no major advances left to be made in the realm of physics. Further advancements lie in the realm of increasingly accurate measurement. However, Lord Kelvin (1824-1907) raised concerns regarding the physics of his era by identifying two significant problems. Which were (Theimer, 1971):

- The theory of luminiferous ether, also known as the wave theory of light, seeks to explain the propagation of light waves in space.
- The limitations of the electromagnetic concept of light in accurately predicting the properties of thermal radiation (Sunyaev & Khatri, 2013).

In the early 1900s, Albert Einstein successfully addressed a fundamental issue by developing a revolutionary concept of relativity. One of the key findings of this theory is the rejection of the concept of ether, along with the notions of absolute space and absolute time. It also challenges the long-held belief that mass and energy are conserved and that nothing can surpass the speed of light. At one point, this theory held immense significance in physics. Max Karl Planck (1858-1947) was able to solve the second problem through the use of quantum mechanical theory. This theory involves the simultaneous existence of both particle and wave nature in all entities. Additionally, it highlights the inherent unpredictability and lack of absolute certainty in the knowledge one can attain (Balta, 2018).

These two theories led to a division in the field of physics. Modern physics covers a wide range of theories that are based on the basic concepts of relativity and quantum mechanics. Quantum mechanics principles were able to effectively clarify the following phenomenon (Wang et al., 2022):

1. The impact of blackbody radiation and the ultraviolet catastrophe.
2. The phenomenon of the photoelectric effect.
3. The point is discrete atomic spectra and the mystery surrounding the existence of atoms.
4. Radioactive decay (Boyer, 2017).

1.1.2. Blackbody Radiation

Objects emit electromagnetic energy as heat regardless of their temperature, with the specific frequencies determined by the temperature. Typically, most of the radiation emitted by an object falls within the infrared range of the electromagnetic spectrum, making it undetectable by the human eye (Prokhorov et al., 2009).

An object that completely absorbs electromagnetic radiation directed toward it, resulting in its visual appearance being black, is known as blackbody. A blackbody, also known as a perfect absorber, is a remarkable radiator that releases radiation across all frequencies. Blackbody radiation is often used to describe the emission of radiation by a blackbody. In practical scenarios, a small aperture located within the heated enclosure, such as a vent in an oven, functions as a blackbody (Cooke & Gallagher, 1980). This occurs because the aperture efficiently captures all radiation that enters (Figure 1.1). The opening to the hollow space within an object might be seen as a precise replica of a blackbody. When light passes through a small opening, it encounters the far wall. Light is reflected multiple times, with some of it being absorbed through the walls of the cavity during each reflection. It is noted that most of all incoming energy is finally assimilated. When the angle of solar radiation changes, as shown in Figure1.1 the energy released from a small hole will eventually reach thermal equilibrium with the surrounding walls. This occurs because of multiple absorptions and re-emissions of the light (Robitaille, 2008).

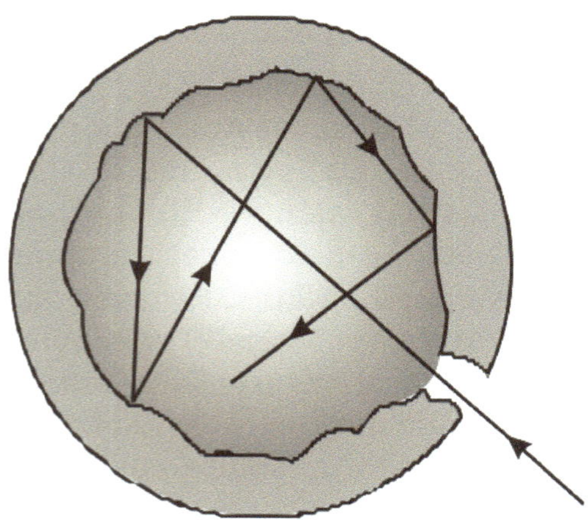

Figure 1.1. *The entrance to the cavity within an object can be considered a close representation of a blackbody.*

Source: P.S. Aithal, Creative Commons License.

1.1.3. Blackbody Radiation Spectrum

To analyze how radiant energy spreads out over different frequencies, it is crucial to maintain a constant temperature for the blackbody. The emissive power of a black body is calculated at various frequencies using an infrared spectrometer. The experimental findings are depicted in Figure 1.2, illustrating the spectrum of blackbody radiation.

Within the spectrum, under specific temperature conditions, the intensity of radiation (energy density) experiences an initial rise as frequency increases. It then reaches a peak around a specific frequency before gradually declining and eventually approaching zero at higher frequencies (Matsumoto & Tomita, 2010).

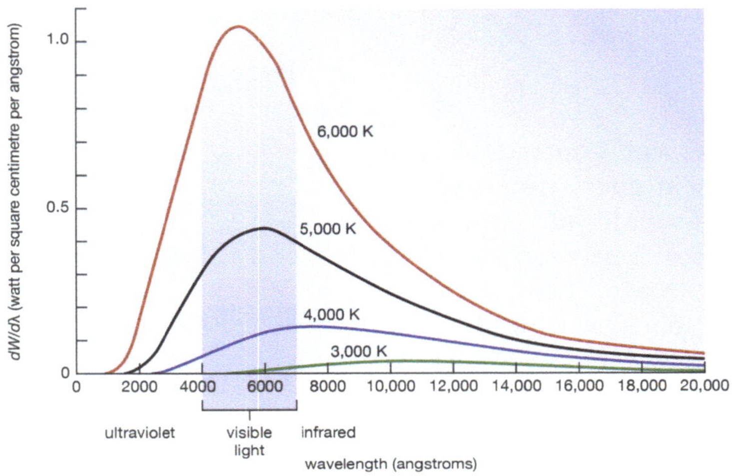

Figure 1.2. *The spectrum of radiations from blackbody*

Source: Ken Stewart, Creative Commons License.

1.1.4. Distribution of Energy in the Spectrum of Blackbody Radiation

By examining the spectrum for blackbody radiation, one can make several important observations:

1. The energy distribution exhibits variations across various frequencies.
2. Temperature and energy density first rise at particular temperatures as the frequency increases reach a peak, and then decline as the frequency continues to grow.
3. As temperature increases, the frequency associated with a maximum energy density (the highest point on the curve) moves to the higher end of the frequency spectrum (Varró, 2006).

1.1.5. The Principles of Blackbody Radiation

Following are the principles of Blackbody Radiation:

1.1.5.1. Stefan's Law

In 1879, the Austrian physicist Josef Stefan conducted an experiment that revealed an intriguing relationship between the total power emitted from a hot solid and its absolute temperature. The total power per unit area was found to be directly related to the temperature raised to one-fourth power. Therefore, the expression for Stefan's law can be written as

surely, it is a revised version of the equation (Johnson, 2012):

$$e_{total} = \int e_f, df = \sigma T^4 \quad (1.1)$$

The sum of all power emitted on the surface of a blackbody at every frequency is denoted by e_{total}. Meanwhile, ef represents the energy emitted per unit area per unit frequency for a blackbody. The absolute temperature for the body is denoted by T, and the Stefan-Boltzmann constant, having a value of 5.67 × 10^{-8} W.m^{-2}.K^{-4}, is symbolized by σ. Non-ideal radiators will have a coefficient, α, that is less than 1 (Reggiani & Alfinito, 2022):

$$e_{total} = \alpha\sigma T^4 \quad (1.2)$$

Where:

- e_{total} (total energy).
- α (proportionality constant).
- σ (Stefan-Boltzmann constant).
- T (absolute temperature) (Kramm & Mölders, 2009).

1.1.5.2. Wein's Distribution Law

Wilhelm Wien put forward a comprehensive explanation of the distribution principle of a blackbody in 1893, which described the observed relationship between λ_{max} and temperature. This principle, often known as Wien's formula, can be expressed as

$$\lambda_{max} T = 2.898 \times 10^{-3}\ m.K \quad (1.3)$$

λ_{max} stands for the wavelength of a blackbody's radiation. While T denotes the absolute temperature. To find the connection between Stefan's law and Wien's displacement law, one must know the surface temperature of the Sun, supposing that behaves as a blackbody.

This is since the human eye has its peak sensitivity at roughly 500 nm:

Here is the calculation along with the equation revised (Wein & Ulbrecht, 1972):

$$T = \frac{2.898 \times 10^{-3}\ m.K}{500 \times 10^{-9}\ m} = 5800\ K \quad (1.4)$$

1.1.5.3. Wein's Exponential Law

Wein in 1893 computed the energy density of radiation inside a blackbody enclosure. His focused on the frequency range spanning from f to $f + df$. The result of his calculation was the determination of $u(f,T)$:

$$u(f,T), df = A f^3 e^{\left\{\frac{\beta f}{T}\right\}}, df \quad (1.5)$$

where A or β represent the constants by Wien. This finding is referred to as Wien's exponential law (Wein, 1990).

1.1.5.4. Limitations of Wein's Law

Experimental evidence indicates that Wein's law is valid in the region of the spectrum characterized by shorter wavelengths and higher frequencies, even when considering blackbodies at high temperatures.

The reason for the progressive decline in the density of energy when longer wavelengths approach the peak value, as shown in Figure 1.3, is not provided (Weinberger, 1991).

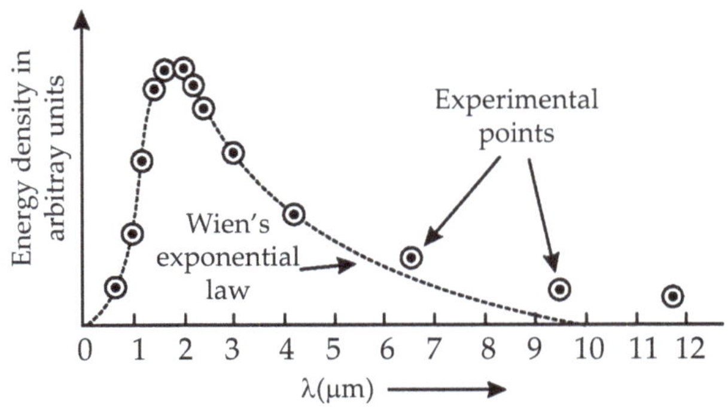

Figure 1.3. *Deviation was observed between Wien's rule and empirical data in a blackbody with a temperature of 1500 K.*

Source: J. S. Aithal, Creative Commons License.

1.1.5.5. Rayleigh-Jeans Law

James Jeans and Lord Rayleigh put forth a proposition regarding the calculation of spectral energy density in each frequency range. And suggested that this density can be determined by considering the density of modes multiplied by $k_B T$,

Below is the equation tidied up and rephrased:

$$u(f\ T)df \quad \frac{f\ k\ T}{\quad} df \qquad (1.6)$$

Where, $u(f,T)$ is spectrum energy density by the frequency (f) and temperature (T); (k_B) represents Boltzmann constant; and c shows the speed of light (Van Vleck, 1924).

1.1.5.6. The Ultraviolet Catastrophe

A blackbody is a theoretical object that can both emit and absorb radiation across all frequencies. The Rayleigh-Jeans law is a result derived from classical physics that describes the relationship between the intensity for blackbody radiation and the frequency at a constant temperature. Rayleigh-Jeans law encounters an issue as the frequency increases, known as the ultraviolet catastrophe. Ultraviolet catastrophe is the divergence observed at high frequencies f (Polzin & Lvov, 2017).

1.1.5.7. The Planck's Radiation Law

In the year 1900, Max Planck, a German scientist, studied the correlation between the temperature of an object. He devised a formula that exhibited remarkable agreement with empirical observations. However, for the formula to hold, he had to assume that the energy from vibrating molecules was quantized, meaning it could only exist in specific, discrete values. It became apparent that the energy was directly related to its frequency of vibration, as it appeared to be quantized into discrete units. Planck's constant, also referred to as h, is a fundamental constant in the field of modern physics. Its numerical value is h = 6.626 × 10^{-34} J.s. Then received the Nobel Prize in 1918, a prestigious honor (Cockell, 1999).

Max Planck formulated the renowned blackbody

Thus, the expression for u(f, T) is:

$$u(f,T)df = \frac{8\pi h f^3}{c^3}\left(\frac{1}{e^{hf/k_B T}-1}\right)df$$

Where $u(f,T)$ the spectral energy density is determined by the frequency f and temperature T; h (Planck's constant); c (speed of light); and k_B (Boltzmann constant).

Whereas k represents Boltzmann's constant, which has a value of 1.380 × 10^{-23} J/K, and h represents Planck's constant, which has a value of 6.626 × 10^{-34} J.s (Dematteis et al., 2022).

If Planck's law can explain how energy is distributed across the entire spectrum in blackbody radiation, it should be possible to derive Wien's and Rayleigh-Jeans laws from it. The reason for this distinction is that Wien's law pertains to the high frequency region, whereas the Rayleigh-Jeans law pertains to the low-frequency region.

The size of the higher frequency $\frac{hf}{k_B T}$ is huge, suggesting the limit of high energy. Here is the way in which it can be stated.

$$e^{\{hf/k_B T\}} \gg 1$$

$$e^{\{hf/k_B T\}} - 1 \approx e^{\{hf/k_B T\}}$$

By replacing the estimate with equation 1.7, one obtains

$$u(f,T)df = \frac{8\pi hf^3}{c^3}\frac{1}{e^{hf/k_B T} - 1}df$$

When one compares the previous equation with Wien's law, the result obtained is

$$A = \frac{8\pi h}{c^3}$$

and

$$\beta = \frac{h}{k_B}$$

Likewise, when it comes to the lower frequency region

f is small, it indicates f is very little $\frac{hf}{k_B T}$ and can be extended $e^{\frac{hf}{k_B T}}$ as a power series (Boleininger et al., 2020).

$$e^{hf/k_{BT}} = 1 + \frac{hf}{k_{BT}} + \frac{1}{2!}\left(\frac{hf}{k_B T}\right)^2 + \frac{1}{3!}\left(\frac{hf}{k_B T}\right)^3 + \ldots$$

Since $\frac{hf}{k_B T}$ size of the terms with higher power is negligible and can be disregarded.

By substituting the aforementioned result into Equation (1.7), one obtains Equation (1.6), known as the Rayleigh-Jeans law (Livi et al., 1987).

1.2. PHOTO-ELECTRIC EFFECT

The photoelectric effect is the phenomenon where electrons are emitted from a metal surface upon exposure to light of a particular frequency. Heinrich Hertz is credited with the discovery of the photoelectric effect in 1887 (Koc, 1957). The photoelectric effect is an intriguing phenomenon that involves the transformation of light energy into electrical energy. Electrons released in the photoelectric effect are commonly referred to as photoelectrons. The flow of these photoelectrons forms what is known as the photoelectric current. The threshold frequency for a metal refers to the minimum frequency of incident radiation required to emit electrons from its surface. The threshold frequency exhibits variation across different metals (Hunter & Alexandrescu, 2005). It has been noted that certain metals, such as zinc, cadmium, and magnesium, exhibit the photoelectric effect exclusively in response to ultraviolet light. Interestingly, several alkali metals, including cesium, lithium, potassium, sodium, and rubidium, demonstrate the photoelectric effect even in the presence of visible light (Poole & Atkins, 1929).

1.2.1. Experimental Study

The experimental setup includes a glass tube that has been evacuated and contains two electrodes (Figure 1.4). Light is directed towards cathode C. The movement of electrons about cathode C toward anode A is observed. The presence of photoelectrons leads to the generation of a photoelectric current, which can be determined by attaching a microammeter to the circuit (Cornelius, 1913). By manipulating the anode potential, one can either increase or decrease the photoelectric current.

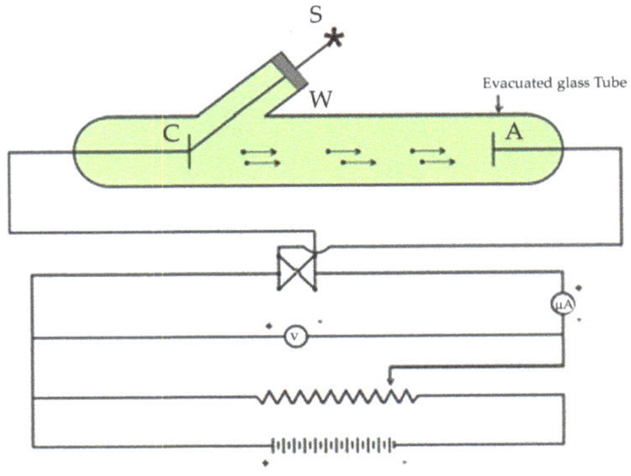

Figure 1.4. *Investigation into the photoelectric effect through experimentation.*

The potential we are referring to is often referred to as the stopping potential, retarding potential, or cut-off potential, and it is denoted as V_s.

The stopping potential is the minimal negative voltage needed on the anode to fully stop the photoelectrons from reaching it, resulting in a complete cessation of current in the circuit (Abdelhady, 2011).

1.2.1.1. Experimental Observations

Following are the experimental observations:

1. The photoelectric effect occurs instantaneously. Photoelectrons are rapidly expelled from the metal surface within a minuscule timeframe of 10–9s following exposure to light irradiation.
2. The photoelectric effect is absent when the frequency of light is below a certain threshold of a particular photosensitive substance.
3. Figure 1.5 (a) demonstrates the direct relationship between the photoelectric current and the intensity of the incident light when considering a specific photosensitive substance or frequency of the reflected radiation (Millikan & Winchester, 1907).
4. One can observe the correlation between the photoelectric current and the applied voltage on each electrode by examining the graph. It is evident that as the potential rises, the photoelectric current initially shows an upward trend. Under specific conditions, the current reaches its highest value for a given potential. The graph demonstrates a plateau in current, indicating a saturation point.
5. Figure 1.6 (b) shows that the stopping potential rises as the frequency of the incident radiation increases.
6. Figure 1.5 (b) illustrates the direct relationship between the frequency of incident radiation and the increase in kinetic energy caused by photoelectrons. The intensity of incoming radiation, as depicted in Figure 1.6 (a), does not affect it (Olpin, 1931).

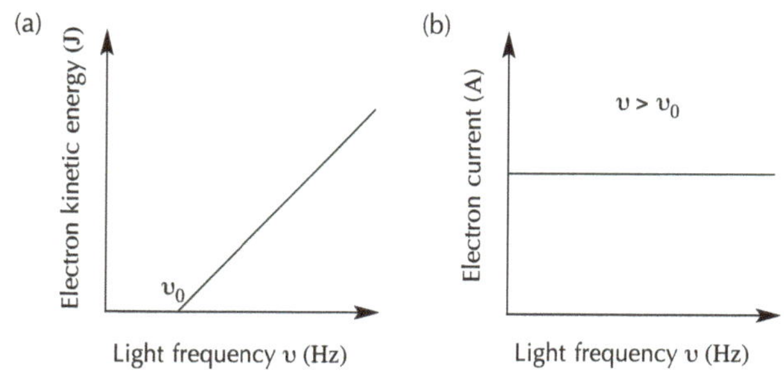

Figure 1.5. *(a) Electron kinetic energy plotted against the frequency of light (b) Graph of electron current verses frequency of incident light.*

Source: Khan Academy, Creative Commons License.

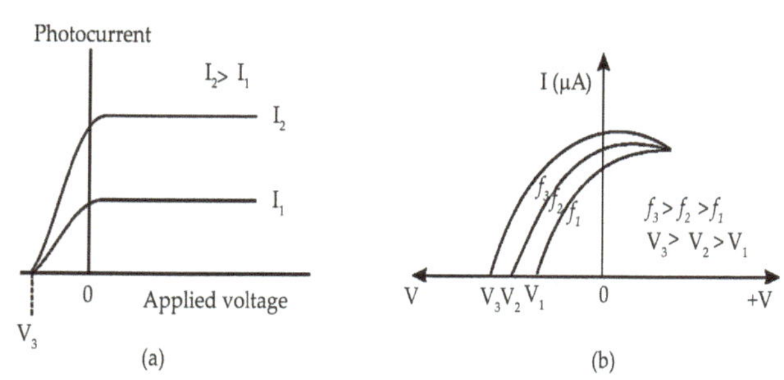

Figure 1.6. *(a) Plot of photoelectric current versus applied potential between electrodes at two different intensity of incident light (b) Plot of photoelectric current versus applied potential between electrodes at three different frequencies of incident light.*

Source: Ravindra H. J., Creative Commons License.

1.2.2. Laws of Photoelectric Effect

Based on the experiment described above, Lenard and Millikan formulated the following laws about the photoelectric effect:

1. The emission rate of photoelectrons via the metal surface has been directly linked to the intensity of the incident light.
2. The intensity of incident light does not affect the highest kinetic energy for emitted photoelectrons.
3. The relationship between the maximum kinetic energy of released photoelectrons and the frequency of incident light is a linear one (Wong et al., 2011).
4. When the frequency drops below a specific threshold, the metal surface does not emit any photoelectrons, irrespective of the light's intensity. The minimum frequency, sometimes referred to as the threshold frequency, differs among various metals.
5. The photoelectric process occurs instantaneously, with no delay between the emission of photoelectrons and the arrival of light.

Traditional physics fell short in its ability to account for the observed phenomena using the theory of electromagnetics (Niaz et al., 2010).

1.2.3. Einstein's Photoelectric Equation

Einstein utilized Planck's Quantum Theory of Light to clarify the experimental observations of the photoelectric phenomenon, which posits that a light beam is composed of individual particles known as photons. He proposed the notion of a work function, which is essentially the minimum energy needed to liberate an electron from a metal surface, denoted as Φ. The measurement of the work function is expressed in electron volts (eV). The work function is influenced by factors, including the characteristics of the metal or surface (Klassen, 2011).

Did you know?

Albert Einstein's explanation of the photoelectric effect, which won him the Nobel Prize in Physics, proposed that light is quantized into particles called photons.

According to Einstein's hypothesis, when a photon with energy hf impacts a metal, the electron promptly assimilates it. When energy absorbed hf exceeds the work function (the energy required to bind the electron), the electron will be released from the metal surface and gain a little momentum to be in a state of motion with a velocity of v.

When an electron absorbs a photon, the energy gets used in two distinct manners.

1. A fraction of the energy through the absorbed photon is used to release the electron onto the exterior of the metal.
2. The energy left is manifested to be the kinetic energy of an electron.

Therefore, one can express as: Energy of incident photon = *Binding energy of electron* + *K.E.* of electron Then,

$$hf = \phi + \tfrac{1}{2} mv^2_{max} \qquad (1.8)$$

Equation 1.8 represents the photoelectric equation created by Einstein, which introduces the concept of a work function for a specific metal. The work process of a metal is defined as the minimum energy needed to release an electron from its surface. Here, $\frac{1}{2} mv^2_{max}$ shows the amount of KE transferred to the electron (Millikan, 1916).

$$K.E. = \frac{1}{2} mv^2_{max} = hf - \phi \qquad (1.9)$$

The work function varies among various metals.

If f_0 denotes the critical frequency, in this context.

$$\phi = hf_0 \qquad (1.10)$$

$$K.E. = \frac{1}{2} mv\ max = hf - \quad = hf - hf_0 = h(f - f_0) \qquad (1.11)$$

Explaining experimental observations through the utilization of Einstein's Photoelectric equation (Kotur, 2023):

1. The photoelectric effect occurs rapidly when an electron collides with a photon, resulting in an instantaneous process.
2. The electron's kinetic energy is determined by the formula $(f - f_0)$ as indicated by equation (1.11). At a frequency $(f - f_0)$, the electron's kinetic energy becomes zero. The frequency f_0 is considered as the threshold frequency. Photoelectrons will not be emitted if the frequency is less than the threshold frequency.
3. When intense light radiation interacts with a metal surface, it results in a greater number of photons interacting with electrons. This, in turn, leads to an increased emission of photoelectrons, thereby causing a rise in the photoelectric current.
4. The energy of the released electrons is directly proportional to the frequency of the incident radiation, as stated by Einstein's equation.
5. As the frequency increases the KE also increases. Consequently, a significant retarding potential is necessary to halt the electrons. When considering the relationship between the mass (m), charge (e), velocity (v), and stopping potential (V) of an emitted electron, it can be expressed as $eV = \frac{1}{2} mv^2$.

Therefore, Einstein effectively elucidated the photoelectric effect by emphasizing the particle-like behavior of light, ultimately in the year 1921 earning a Noble Prize (Pais, 1979).

1.3. COMPTON EFFECT

In 1916, Einstein expanded his understanding of photons and suggested that when light comes into contact with matter, the energy and linear momentum of the photon are transferred in discrete amounts. This discovery reaffirmed the principles of the conservation of energy and momentum. The magnitude sent with photon for frequency f can be determined using the following equation (Everett et al., 1977):

$$p = hf / c = h / \lambda \quad (1.12)$$

In 1923, Arthur Compton conducted a groundbreaking experiment at Washington University that provided evidence in favor of Einstein's theories. Compton conducted an experiment where he directed a beam of X-rays onto a graphite target and analyzed the scattered X-rays' wavelengths and intensities in different directions. The phenomenon that has been observed is known as the Compton effect. He was awarded the Nobel Prize in 1927 (Glover, 1982).

When X-rays with a specific frequency encounter a material with a low atomic number, such as carbon, it undergoes a frequency change due to scattering. The scattered beam consists of two different wavelengths. Besides the anticipated incident wavelength, there is also a portion with a longer wavelength. This phenomenon is often referred to as the Compton effect. The alteration in the wavelength ($\Delta\lambda$) from the initial wavelength (λ) to the resulting wavelength of the scattered X-rays (λ') is known as the Compton shift (Baranov et al., 1968).

Explanation: This theory of electromagnetics was unable to explain the Compton shift phenomenon. However, Compton effectively utilized the Quantum Theory of Radiation to clarify this phenomenon. The principles of conservation are upheld for both momentum and energy. During this collision between a photon and an electron, a fraction of the photon's energy goes to the electron. Consequently, the X-ray undergoes a decrease in energy, resulting in a lower frequency (Arutyunian & Tumanian, 1963).

1.3.1. Expression for Compton Shift

A photon, carrying energy *hf* ' or momentum *hf/c*, collides with an electron at rest. The electron initially possesses zero momentum; hence its initial energy is completely derived from its rest mass, which is equal to moc^2 (Prange, 1958). The photon, using energy *hf* and momentum *hf'/c*, undergoes scattering and thereafter proceeds along an angle θ, as seen in Figure 1.7: Electron gets an energy level of mv or alters its course by an angle θ concerning its original path. The energy generated by the recoil electron is determined by mc^2 (Radyushkin and Weiss, 2001).

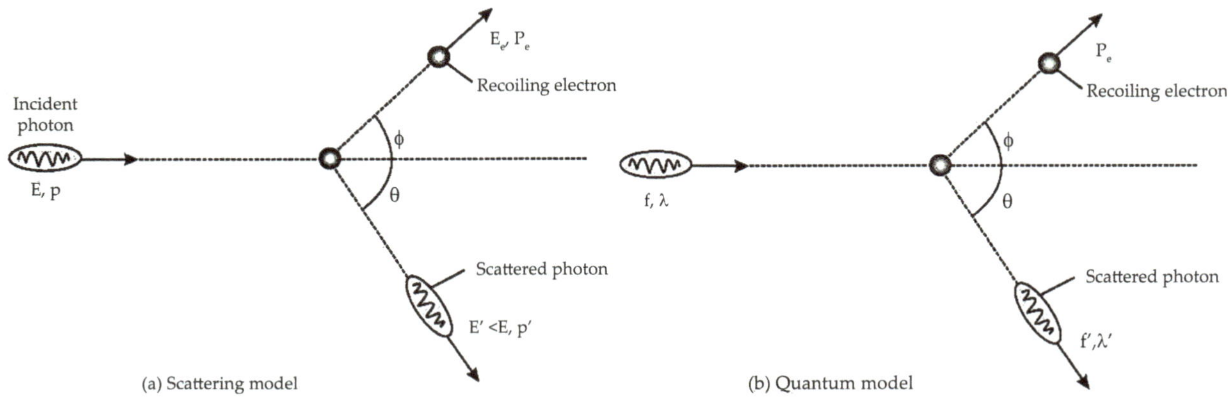

Figure 1.7. *Scattering of photon by electron of an atom.*

Source: Ravindra H. J., Creative Commons License.

Based on the fundamental principle of energy conservation,

$$hf + m_o c^2 = hf' + mc^2 \quad (1.13)$$

By considering both the x and y parts of momentum and applying the principle to conserve momentum, one can derive (Dirac, 1926):

$$hf/c + 0 = \left(hf'/c\right)\cos\theta + mv\cos\varphi \quad (1.14)$$

and

$$0 = \left(hf'/c\right)\sin\theta - mv\sin\varphi \quad (1.15)$$

The symbol 'm' is used to represent the mass of an electron in motion, which is related to its mass at rest. According to the Einstein Relativistic equation,

$$m = \frac{m_0}{\sqrt{1-\left(\frac{v_2}{c^2}\right)}} \quad (1.16)$$

Through Equation. (1.14) and (1.15) one obtained

$$hf - hf'\cos\theta = mvc\cos\varphi \quad (1.17)$$

$$hf'\sin\theta = mvc\sin\varphi \quad (1.18)$$

Operating squaring and addition on equations (1.17) and (1.18),

$$H^2 f^2 + h^2 f'^2 - 2h^2 ff'\cos\theta = m^2 v^2 c^2 \quad (1.19)$$

Through Equation. (1.13),

$$hf - hf' + m_0c^2 = mc^2 \quad (1.20)$$

Squaring the Equation. (1.20) provides

$$\left(hf - hf'\right)^2 + m_0^2c^4 + 2m_0c^2h\left(f - f'\right) = m^2c^4$$

or

$$h^2f^2 - h^2f'2 - 2h^2ff' + m_0^2c^4 + 2m_0c^2h\left(f - f'\right) = m^2c^4 \quad (1.21)$$

Doing the step of subtraction of Equation (1.19) from Equation (1.21) one get (DuMond & Kirkpatrick, 1937),

$$m^2c^2\left(c^2 - v^2\right) = -2h^2ff' + 2m_0c^2h\left(f - f'\right) + 2h^2ff'\cos\theta + m_0^2c^4$$

$$m^2c^2\left(c^2 - v^2\right) = -2h^2ff'\,(1 - \cos\theta) + 2m_0c^2h\left(f - f'\right) + m_0^2c^4 \quad (1.22)$$

From Equation. (1.16) one get,

$$m^2 = m_0^2c^2 / \left(c^2 - v^2\right)$$

or

$$m^2\left(c^2 - v^2\right) = m_0^2c^2 \quad (1.23)$$

Utilizing Equation. (1.23) in Equation. (1.22), one get

$$M_0^2c^4 = -2h^2ff'\,(1 - \cos\theta) + 2h\left(f - f'\right)m_0c^2 + m_0^2c^4$$

or

$$2h\left(f - f'\right)m_0c^2 = 2h^2ff'\,(1 - \cos\theta)$$

or

$$\left(f - f'\right)/ff' = \left(h/m_0c^2\right)(1 - \cos\theta)$$

or $1/f' - 1/f = \left(h/m_0c^2\right)(1 - \cos\theta)$

or $c/f' - c/f = (h/m_0c)(1 - \cos\theta)$

or $\lambda' - \lambda = (h/m_0c)(1 - \cos\theta)$

or $\lambda' = \lambda + (h/m_0c)(1 - \cos\theta) \quad (1.24)$

Hence, the alteration in wavelength =

$$\Delta\lambda = \lambda' - \lambda = (h/m_0c)(1 - \cos\theta) \quad (1.25)$$

Equation (1.25) represent (McGovern et al., 2013):

1. The Compton shift Δλ remains unaffected by the wavelength.
2. The Compton shift Δλ remains unaffected by the composition of the scattering substance.
3. The Compton shift Δλ is solely determined by the angle for scattering θ.
4. The wavelength of scattered light is constantly greater than the wavelength in incident light.

Three instances can be explored based on the angle of scattered light:

- **Case (1):** Where θ = 0°: Δλ = 0, i.e., having no scattered photon throughout this incident.
- **Case (2):** Where θ = 90°: Δλ = h/m0c = 0.02426 Å, It is called Compton wavelength.
- **Case (3):** Where θ = 180°: Δλ = 2 h/m0c = 0.04852 Å, Is double of Compton wavelength (Krajewska and Kamiński, 2012).

1.3.2. Experimental Confirmation of Compton Shift

Monochromatic X-rays of a specific wavelength (λ) are directed at a scattering substance, such as a tiny block of carbon. The Bragg spectrometer receives the dispersed X-rays and measures their wavelength (Figure 1.8) (Wolf et al., 2001). The spectrometer can oscillate unrestrictedly in a curved path around the scatterer. The wavelength for the X-rays that have been dispersed is determined by measuring the scattering angle at various values. Figure 1.9 displays the experimental findings acquired by Compton. Within the scattered radiation, there is not only the original wavelength (λ) but also a longer wavelength (λ') present. The Compton shift Δλ seems to exhibit variation with the angle of observation of the scattered photons (Drechsel et al., 2003).

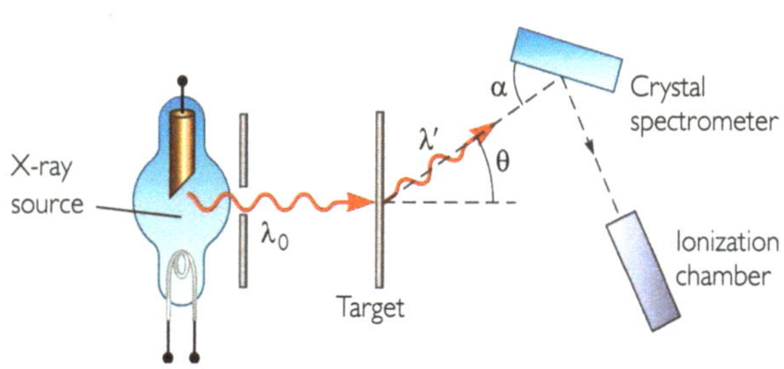

Figure 1.8. *Illustration of the setup of the Compton experiment.*

Source: Science ABC, Creative Commons License.

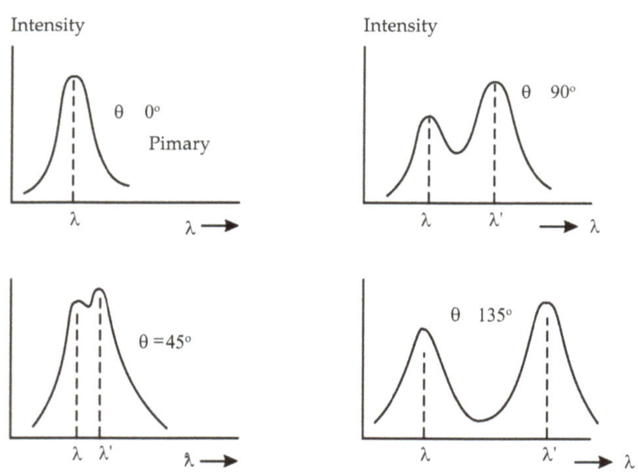

Figure 1.9. *Estimating the X-ray intensity as a function of the wavelength for Compton scattering at different angles (0°, 45°, 90°, or 135°).*

Source: Mohamed Boubendir, Creative Commons License.

1.4. WAVE PARTICLE DUALISM

The wave theory of light provides a clear and concise explanation for several optical phenomena, including interference, diffraction, reflection, polarization, and refraction. However, the wave theory of light was inadequate in elucidating the phenomena of the photoelectric effect, blackbody radiation, and Compton effect (Chiatti, 2020). The quantum concept of light provides a straightforward explanation for these events. This demonstrates that light can manifest in both forms, providing evidence for both wave theory and quantum theory. The dual character of light is attributed to its intricate characteristics. Photons, the particles that carry energy, demonstrate diffraction effects, indicating their wave-like nature. However, this observation leads to a significant conclusion that light has dual characteristics, specifically, it simultaneously functions as a particle and a wave (Menzel et al., 2012).

1.4.1. de-Broglie Hypothesis

In 1923, Louis de-Broglie put forward the idea that matter, like radiation, possesses a dual nature. Under certain conditions, matter composed of discrete particles such as atoms, protons, and electrons can display wavelike characteristics. In his doctoral thesis, de-Broglie proposed that the dual nature of photons, exhibiting both wave and particle features, could extend to all forms of matter. In 1929, he received the Nobel Prize for his remarkable discovery of the wave properties of moving particles.

And put forth a series of parallels to support his hypothesis (Adamua et al., 2024):

1. **Nature Loves Symmetry:** Nature presents itself in two distinct forms: matter and radiation. It is understood that radiation exhibits a dual nature, which suggests that matter may likewise exhibit a dual nature. Similarly, it is seen as an electric field that is shifting and gives rise to a magnetic field, according to the principle of symmetry.
2. **The Striking Similarity Between Mechanics and Optics:** In mechanics, the principle of least action states that a moving particle will always choose the path that minimizes the action. Light always takes the shortest path, according to the optics law of Fermat.
3. **Bohr's Atomic Structure Theory:** Bohr's hypothesis argues that the stable states of electrons in an atom are determined by an integer rule. Wave motion is the sole phenomenon in physics that involves integers. Therefore, de-Broglie postulated that electrons may likewise exhibit periodic behavior like waves.

These similarities indicate that an electron like other material particles must demonstrate both wave-like and particle-like characteristics. The term "matter waves" refers to waves that are connected to a physical particle (Frieden & Soffer, 2009).

Remember

Blackbody radiation demonstrates that the spectrum of emitted radiation depends only on the temperature of the object, not its composition.

1.4.2. de-Broglie's Wavelength

The de-Broglie matter wave expression may be derived by combining Einstein's and Max Planck's mass energy with energy equation.

Einstein claimed that a photon traveling at the speed of light, c, has a mass-energy relationship of

$$\pm = \quad {}^{2} \tag{1.26}$$

Max Planck proposed that the energy of a photon with a frequency f can be determined using the following equation:

$$E = hf = hc/\lambda \tag{1.27}$$

Comparing Equation (1.26) and (1.27), $mc^2 = hc/\lambda$

or $\lambda = h/mc$ (1.28)

Equation (1.28) is the formula for the wavelength for a photon wave in a medium if the photon is traveling at the speed of light (c) (Liu et al., 2008).

Similarly, the de-Broglie wavelength of a material particles with mass m or velocity v is given by

$$\lambda = h/mv \tag{1.29}$$

When a particle of material moves in motion and demonstrates $E = ½ mv^2$, the de-Broglie wavelength might be expressed as

$$\lambda = \frac{h}{mv} = \frac{h}{\sqrt{2mE}} \tag{1.30}$$

1.4.3. de-Broglie's Wavelength-Extension to Electron Particle

Let's examine the case of an electron with a rest mass of m_0 and a charge of e that is accelerated by a voltage of V volts, starting at rest and reaching a velocity of v. Next, one begins to write:

$$½ m_o v^2 = eV$$

or $v = \sqrt{\frac{2eV}{m_0}}$ (1.31)

It is also known that the wavelength for matter waves is $\lambda = h/m_0v$ (1.32). Utilizing Equation (1.31) and (1.32), one writes:

$$\lambda = \frac{h}{\sqrt{2m_0eV}} \quad (1.33)$$

Placing values m = 9.11 × 10^{-31} kg; h = 6.62 × 10^{-34} J/s and e = 1.602 × 10^{-19} C, we get Equation (1.33) as

$$\lambda = \frac{12.26}{\sqrt{V}} . \mathring{A} \quad (1.34)$$

When the accelerating voltage is set at 100 volts, the wavelength of the electron is calculated to be 1.226 Å. Given the similarity in wavelength between de-Broglie waves and X-rays, it is reasonable to expect that de-Broglie waves can undergo diffraction by a crystal, like X-rays (Katzer et al., 2024).

1.4.4. Davison and Germer Experiment

C. J. Davison and L.H. Germer in 1927 from the Bell Telephone Laboratory of New York experimented to validate de-Broglie theory concerning the presence of matter waves. This experiment achieved a noteworthy milestone in the realm of physics (Morrison, 1986).

1.4.4.1. Experimental Setup

Figure 1.10 displays the experimental setup. Electrons are produced by heating a filament via a low-voltage battery. Electrons are accelerated through the application of a speeding voltage; V. A movable detector is used to detect the collected beam. The setup is located within a vacuum chamber.

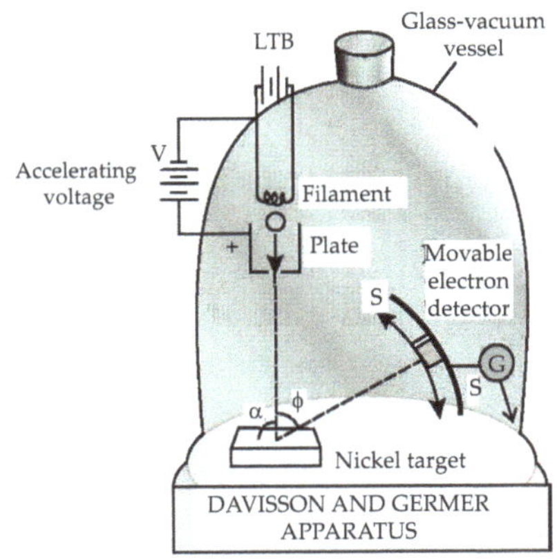

Figure 1.10. *A picture illustrating the apparatus of Davison-Germer.*

1.4.4.1. Experimental Procedure

An electron beam is directed toward the surface of the crystal. The reflected beam is captured by a detector and adjusted on a circular scale S. To determine the electron beam intensity, a galvanometer G is used.

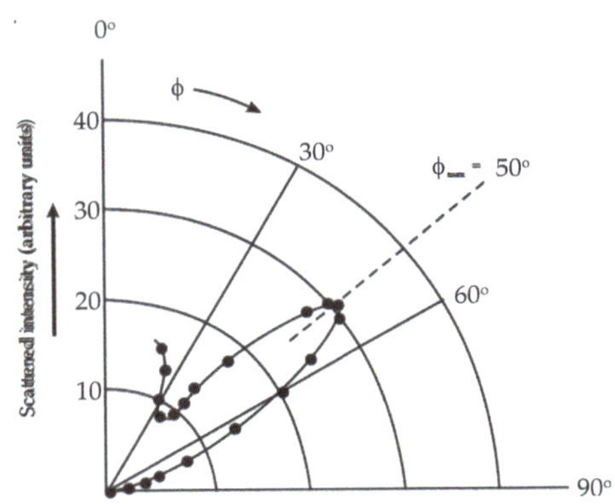

Figure 1.11. *Graphing the intensity of the diffracted beam as a function of the angle between the incident or diffracted beams at a voltage of 54 V.*

A graph is created by plotting the deflection of the galvanometer against the angle formed between the incident and beam that enters the detector. Figure 1.11 displays a standard graph illustrating the potential for acceleration of 54 V (Germer, 1925).

1.4.4.2. Experimental Result

The graph maintains a consistent smoothness until the accelerating voltage reaches 44 volts. At this point, an intriguing spike emerges on the graph. As the accelerating voltage continues to rise, the length of this spike gradually increases. Eventually, at an angle of 50°, the accelerating voltage reaches to its peak of 54 volts and decreases until it eventually disappears at 68 volts (Calbick, 1963).

1.4.4.3. Inference

One possible explanation for the presence of a distinct peak at 50o, when an electron is accelerated by 54 volts, is the constructive interference for electron waves. These waves are scattered in that particular direction by the frequently spaced parallel surfaces within the Nickel crystal. The measured wavelength is determined by equation:

$$\lambda = \frac{12.26}{\sqrt{54}}.\ \text{Å} = 1.66\ \text{Å}$$

The experiment showed that the light beam diffracted at 50o co-latitude matches an electron beam's usual glancing angle $\theta = 65°$ around the crystal plane. From X-ray diffraction studies, it was determined that the interplanar distance for nickel is 0.91 Å. In this case, one considers the order for diffraction to be 1. To calculate it, one can apply Bragg's equation: $n\lambda = 2d\sin(\theta)$, where n represents the order of diffraction. After simplifying the result is $\lambda = 2 \times 0.91$ Å $\times \sin 50° = 1.65$ Å. Therefore, the value closely aligns to the theoretical value. It is evident that a stream of electrons exhibits similar behavior to X-rays, experiencing diffraction when encountering reflecting surfaces, indicating its wave-like properties (Deutsch, 1958).

1.5. MATTER WAVES AND PROPERTIES

In contrast to electromagnetic waves, matter wave exhibits the following characteristics:

1. The relationship between the wavelength of a matter wave and the mass of a particle is inversely correlated. Therefore, the wavelength of a particle will become shorter if it has a larger mass, vice versa.
2. The relationship between the wavelength of a matter wave and the velocity of the particle is inversely proportional. Therefore, as the velocity of a particle increases, the wavelength decreases, and vice versa.
3. De-Broglie (amplitude) wave changes depending upon probability of detecting a particle at a specific location. Therefore, the matter waves are referred to as probability waves.
4. If a particle is moving, it relates to two distinct velocities. The mechanical motion of particles refers to the physical movement ($v_{particle}$) while the motion of linked matter waves refers to the movement of the waves (v_{phase}). The connection between these two is shown by equation $v_{phase} = c^2/v_{particle}$. The velocity of transmission of matter waves v_{phase} < than c. It is interesting to note that matter waves can surpass the speed of light.

1.5.1. Particle Velocity, Phase Velocity, and Group Velocity

De-Broglie proposed that particles of matter, such as electrons and protons, can be understood as a collection of wave packets or as a wave. Every individual wave travel at a specific speed known as phase velocity and wave velocity. Nevertheless, in the case of a disturbance consisting of multiple elemental waves, all having slightly different velocities, the resulting velocity will exhibit periodic behavior. The rate at which this periodicity moves forward is referred to as group velocity (Arbab et al., 2012).

1.5.1.1. Phase Velocity

Wave propagating on the positive x-direction has the equation

$$y = A \sin (\omega t - kx) \qquad (1.35)$$

The variables in the equation are specified as follows: y indicates the displacement in the vertical direction at a certain time t, ω indicates the rate of rotation, k represents the propagation constant, and x indicates the displacement in the horizontal direction at the same time t. The given equation depicts the state of the vibrating particle as (ωt-kx). Particles in the same phase exhibit stationary behavior, as indicated by a zero shift (ωt-kx) over time, showing no temporal variation,

$$\frac{d\left(w_0 t - kx\right)}{dt} = 0$$

Or

$$w - k\frac{dx}{dt} = 0$$

Or

$$\frac{dx}{dt} = \frac{w}{k}$$

Phase velocity refers to the rate at which the change in phase of a vibrating particle progresses as the wave goes forward. Within this context, the derivative of x with respect to t, denoted by $\frac{dx}{dt}$, is regarded as the phase velocity (Pialucha et al., 1989).

$$v_{phase} = \frac{w}{k}$$

1.5.1.2. Group Velocity

Definition: The group velocity refers to the speed at which the wave packet, created by combining multiple traveling waves with slightly different frequencies, is transported (Figure 1.12).

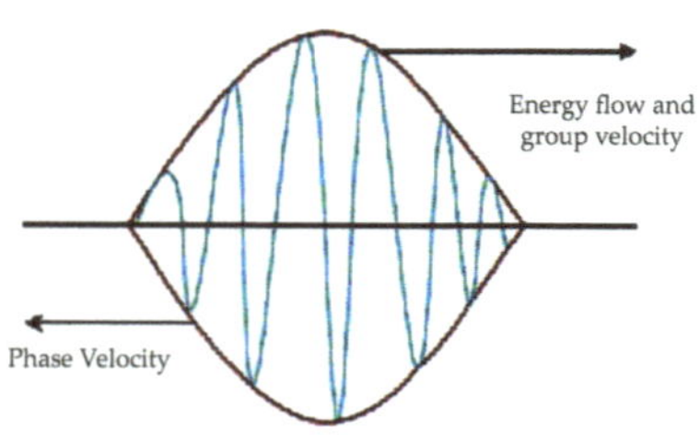

Figure 1.12. *Illustration upon group velocity.*

Source: Go Photonics, Creative Commons License.

Let's examine two traveling waves, y_1, and y_2, which have the similar amplitude but different wavelengths. These waves can be expressed as:

$$y_1 = A\ \sin(\omega_1 t - k_1 x)$$

$$y_2 = A\ \sin(\omega_2 t - k_2\ x)$$

The equation for calculating the final displacement y resulting from a superposition of the two waves is as follows:

$$y = y1 + y2$$

$$y = A\left[\ sin\left(\omega_1 t - k_1 x\right) + sin\left(\omega_2 t - k_2 x\right)\right]$$

Utilizing the identity

$$\sin\alpha + \sin b = 2\cos\left(\frac{a-b}{2}\right)\sin\left(\frac{a+b}{2}\right)$$

one gets

$$y = 2A\cos\left(\frac{\left(w_1 - w_2\right)t}{2} - \frac{\left(k_1 - k_2\right)x}{2}\right)\sin\left(\frac{\left(w_1 + w_2\right)t}{2} - \frac{\left(k_1 + k_2\right)x}{2}\right)$$

$$y = 2A\cos\left(\frac{\Delta wt}{2} - \frac{\Delta kx}{2}\right)\sin(wt - kx) \qquad (1.36)$$

Whereas

$\omega = \omega_1 + \omega_2\ /2$, $k = k_1 + k_2/2$, $\omega = (\omega_1 - \omega_2)$, $k = k_1 + k_2/2$

Comparing Equation (1.35) with Equation (1.36). In equation (1.35), 'A' represents the amplitude, whereas in equation (1.36), the amplitude is

$$2A\cos\left(\frac{\Delta wt - \Delta kx}{2}\right)$$

The amplitude of the resultant wave above remains constant. The rate at which the amplitude change is transmitted within the resultant waveform is referred to as the group velocity. And it can be determined using the following equation (Whitham, 1960):

$$v_{group} = \Delta\omega/\Delta k \qquad (1.37)$$

1.5.2. The Relationship between Phase Velocity and Group Velocity

Formula for calculating the group velocity can be derived as follows:

$$V_{group} = d\omega / dk \qquad (1.38)$$

The formula of phase velocity can be derived as follows:

$$V_{phase} = \omega/k \qquad (1.39)$$

In this context, ω represents angular frequency, while k denotes the wave vector.

The equation (1.39) can be expressed as $\omega = kV_{phase}$

Thus,

$$v_{group} = \frac{dw}{dk} = \frac{d}{dk}\left(kv_{phase}\right) = v_{phase} + k\frac{dv_{phase}}{dk}$$

$$v_{group} = v_{phase} + k\left(\frac{dv_{phase}}{d\lambda}\right)\frac{d\lambda}{dk} \qquad (1.40)$$

But, one knows that value of $k = 2\pi/\lambda$

Differentiating one get,

$$dk / d\lambda = -2\pi / \lambda^2,$$

or $d\lambda / dk = -\lambda^2 / 2\pi$

Thus, can write

$$k\left(d\lambda / dk\right) = \left(2\pi / \lambda\right)\left(-\lambda^2 / 2\pi\right) = -\lambda$$

Putting the above Equation in Equation (1.40), one get,

$$v_{group} = v_{phase} - \lambda\left(\frac{dv_{phase}}{d\lambda}\right) \qquad (1.41)$$

The link between both velocities may be deduced from Equation (1.41) (Wear, 2000).

1.5.3. Relation between Group Velocity and Particle Velocity

The formula for calculating group velocity is provided by

$$V_{group} = d\omega / dk \qquad (1.42)$$

Bu

$$\omega = 2\pi f = 2\pi\left(E / h\right) \qquad (1.43)$$

$$d\omega = \left(2\pi / h\right) dE \qquad (1.44)$$

Also we have

$$k = 2\pi / \lambda = 2\pi\left(p / h\right) \qquad (1.45)$$

$$dk = \left(2\pi / h\right) dp \qquad (1.46)$$

After dividing Equation (1.44) by Equation (1.46), one obtains

$$d\omega / dk = dE / dp \qquad (1.47)$$

Knowing the kinetic energy of particle is $E = p^{2/}2m$, where p stands for the momentum. Then,

$$dE / dp = 2p / 2m = p / m \qquad (1.48)$$

Utilizing Equation (1.48) in Equation (1.47) one gets, $d\omega / dk = p / m$

But $p = mV_{particle}$ where $V_{particle}$ shows the velocity of the particle.

Then,

$$d\omega / dk = mV_{particle} / m = V_{particle} \qquad (1.49)$$

through Equation (1.42) and (1.49), one gets,

$$V_{group} = V_{particle} \qquad (1.50)$$

Nevertheless, it is widely accepted that the movement of a particle can be determined by employing the equation E = $p_2/2m$, when p denotes the particle's momentum (Ruoff, 1967).

1.5.4. Deriving the de-Broglie Wavelength through the concept of Group Velocity

The formula for the velocity of

$$V_{group} = d\omega / dk \quad (1.51)$$

But $\omega = 2\pi f \text{ and } k = 2\pi/\lambda$

Then $d\omega = 2\pi df \text{ and } dk = 2\pi d(1/\lambda)$

Then Equation (1.51) becomes

$$V_{group} = d\omega / dk = df / d(1/\lambda) \quad (1.52)$$

The particle velocity can be seen as equivalent to the group velocity.

$$v = df / d(1/\lambda) \quad \text{or} \quad d(1/\lambda) = df / v \quad (1.53)$$

By utilizing the concept of total energy for a particle, one could write

$$hf = \tfrac{1}{2}mv^2 + V \quad (1.54)$$

By considering a particle in motion in a field of constant potential V, one can derive the resulting equation by distinguishing Equation 1.54:

$$h\,df = mv.dv$$

or

$$df / v = (m/h)dv \quad (1.55)$$

From Equation (1.53) and (1.55), we get,

$$d(1/\lambda) = (m/h)dv$$

Integrating,

$$1/\lambda = (m/h)v + constant \quad (1.56)$$

Consider the momentum of a particle as p, which can be expressed as p = mv. Furthermore, when substituting the intrinsic value of integration as zero, one obtains,

$$1/\lambda = p/h \text{ or } \quad \lambda = h/p \quad (1.57)$$

The above Equation (1.57) is the de-Broglie's equation (Khadka, 2022).

1.5.5. Exploring the Connection among Phase Velocity, Group Velocity, and the Speed of Light

The expression for phase velocity is given by $V_{phase} = \omega / k$

$$\omega = 2\pi f = 2\pi \frac{E}{h} \quad (1.58)$$

Also

$$k = \frac{2\pi}{\lambda} = 2\pi \frac{p}{h} \quad (1.59)$$

Then

$$v_{phase} = \frac{w}{k} = \frac{E}{p} = \frac{mc^2}{mv_{phase}} = \frac{c^2}{v_{phase}}$$

But, from Equation (1.50) (Stenner et al., 2003), $v_{group} = v_{particle}$ Utilizing Equation (1.60) and (1.50), one can write as

$$V_{phase} = V_{group} = c^2 \quad (1.61)$$

ACTIVITY 1.1.

Conduct a simulation or experiment to demonstrate the photoelectric effect using a light source, a metal plate, and a voltmeter. Measure how varying the intensity and frequency of light affects the emission of electrons and their kinetic energy, thereby illustrating the principles described by Einstein's photoelectric equation.

SUMMARY

This chapter delves into the foundational concepts of modern physics, commencing with an overview of its significance in advancing our understanding of the universe. The discussion begins with blackbody radiation, a critical phenomenon that characterizes the emission of radiation by objects based on their temperature. The chapter examines the blackbody radiation spectrum, elucidating the energy distribution across various wavelengths and the underlying principles governing this distribution.

After discussing blackbody radiation, the chapter transitions to the experimental study of the photoelectric effect. This section explains the laws that describe the behavior of electrons when illuminated by light, resulting in their emission. Einstein's photoelectric equation is introduced, providing a theoretical framework that revolutionized the understanding of light and its interaction with matter.

The exploration of modern physics continues with an analysis of the Compton effect, in which the chapter provides a detailed derivation of the expression for the Compton shift and discusses its experimental validation. This effect supports the particle-like behavior of light, posing a significant challenge to classical wave theories.

Following this, the chapter introduces the de Broglie hypothesis, which postulates that particle, such as electrons, exhibit wave-like properties. The concept of de Broglie wavelength is explained, including its application to electron particles. This wave-particle duality is experimentally validated through the Davisson-Germer experiment, which confirmed the wave nature of electrons and provided critical evidence for quantum mechanics.

The chapter also examines the relationship between phase velocity and group velocity, elucidating how these concepts pertain to the behavior of particles and waves. It presents the derivation of the de Broglie wavelength through the concept of group velocity, offering deeper insights into the nature of matter waves.

Finally, the chapter explores the interrelation among phase velocity, group velocity, and the speed of light. This section delves into the theoretical implications of these relationships, enhancing the understanding of wave-particle duality and its role in modern physics.

REVIEW QUESTIONS

1. What is blackbody radiation, and why is it significant in the study of modern physics?
2. Describe the blackbody radiation spectrum and explain how energy distribution varies across different wavelengths.
3. What are the key laws governing the photoelectric effect, and how does Einstein's photoelectric equation explain this phenomenon?
4. Derive the expression for the Compton shift and discuss the experimental evidence that confirms this effect.
5. Explain the de Broglie hypothesis and how it contributes to our understanding of wave-particle duality.
6. What is de Broglie wavelength, and how is it applied to electron particles? Provide an example.
7. Describe the Davisson-Germer experiment and its significance in validating the wave nature of electrons.
8. Discuss the relationship between phase velocity, group velocity, and particle velocity, and explain how these concepts relate to the de Broglie wavelength and the speed of light.

MULTIPLE CHOICE QUESTION

1. Which phenomenon describes the emission of radiation by an object based on its temperature?

a. Photoelectric effect
b. Compton effect
c. Blackbody radiation
d. de Broglie hypothesis

2. What is the significance of the blackbody radiation spectrum?

a. It explains the behavior of light as particles.
b. It describes the energy distribution of radiation across various wavelengths.
c. It validates the wave nature of electrons.
d. It provides a framework for the Compton effect.

3. Einstein's photoelectric equation explains the photoelectric effect by relating the energy of the emitted electrons to which factor?

a. The frequency of the incident light
b. The wavelength of the incident light
c. The temperature of the emitting surface
d. The angle of incidence of the light

4. **The Compton effect demonstrates which property of light?**
   a. Wave nature
   b. Particle nature
   c. Dual nature
   d. Reflective nature
5. **According to the de Broglie hypothesis, particles such as electrons exhibit which characteristic?**
   a. Only wave properties
   b. Only particle properties
   c. Both wave and particle properties
   d. Neither wave nor particle properties
6. **What experimental evidence supports the wave nature of electrons?**
   a. Photoelectric effect
   b. Compton shift
   c. Davisson-Germer experiment
   d. Blackbody radiation
7. **Which of the following correctly describes the relationship between phase velocity (v_p), group velocity (v_g), and the speed of light (c)?**
   a. v_p = v_g = c
   b. v_p > v_g = c
   c. v_p < v_g < c
   d. v_p * v_g = c^2

Answers to Multiple Questions

1. (c); 2. (b); 3. (a); 4. (b); 5. (c); 6. (c); 7. (d)

REFERENCES

1. Abdelhady, S. (2011). Comments on Einstein's explanation of electrons, photons, and the photoelectric effect. *Applied Physics Research, 3*(2), 230–240.
2. Adamua, I., Hassan, A., Maina, M., Kole, A. M., & Dawud, J. A. (2024). Review on de Broglie stationary wave hypothesis from classical mechanics. *NIPES Journal of Science and Technology Research, 6*(1), 654–660.
3. Arbab, A. I., Widatallah, H. M., & Khalafalla, M. A. H. (2012). Toward a complete wave-particle duality: Do matter waves have inertia? *Advances in Natural Science, 5*(3), 32–34.
4. Arutyunian, F. R., & Tumanian, V. A. (1963). The Compton effect on relativistic electrons and the possibility of obtaining high energy beams. *Physics Letters, 4*(3), 176–178.

5. Balta, N. (2018). High school teachers' understanding of blackbody radiation. *International Journal of Science and Mathematics Education, 16*, 23–43.
6. Baranov, P. S., Fil'kov, L. V., & Sokol, G. A. (1968). Proton Compton effect. *Fortschritte der Physik, 16*(11, 12), 595–634.
7. Boleininger, M., Swinburne, T. D., Dupuy, L., & Dudarev, S. L. (2020). Ultraviolet catastrophe of a fluctuating curved dislocation line. *Physical Review Research, 2*(3), 032033.
8. Boyer, T. H. (2012). The blackbody radiation spectrum follows from zero-point radiation and the structure of relativistic spacetime in classical physics. *Foundations of Physics, 42*, 595–614.
9. Boyer, T. H. (2017). Scaling, scattering, and blackbody radiation in classical physics. *European Journal of Physics, 38*(4), 045101.
10. Calbick, C. J. (1963). The discovery of electron diffraction by Davisson and Germer. *The Physics Teacher, 1*(2), 63–91.
11. Chandler, D. (1987). *Introduction to Modern Statistical Mechanics.* Oxford University Press.
12. Chiatti, L. (2020). Bit from qubit: A hypothesis on wave-particle dualism and fundamental interactions. *Information, 11*(12), 571.
13. Cockell, C. S. (1999). Crises and extinction in the fossil record—A role for ultraviolet radiation? *Paleobiology, 25*(2), 212–225.
14. Cooke, W. E., & Gallagher, T. F. (1980). Effects of blackbody radiation on highly excited atoms. *Physical Review A, 21*(2), 588–592.
15. Cornelius, D. W. (1913). The velocity of electrons in the photoelectric effect, as a function of the wavelengths of the light. *Physical Review, 1*(1), 16–25.
16. Dematteis, G., Polzin, K., & Lvov, Y. V. (2022). On the origins of the oceanic ultraviolet catastrophe. *Journal of Physical Oceanography, 52*(4), 597–616.
17. Deutsch, M. (1958). Evidence and inference in nuclear research. *Daedalus, 87*(4), 88–98.
18. Dirac, P. A. M. (1926). Relativity quantum mechanics with an application to Compton scattering. *Proceedings of the Royal Society of London. Series A, Containing Papers of a Mathematical and Physical Character, 111*(758), 405–423.
19. Drechsel, D., Pasquini, B., & Vanderhaeghen, M. (2003). Dispersion relations in real and virtual Compton scattering. *Physics Reports, 378*(2, 3), 99–205.
20. DuMond, J. W., & Kirkpatrick, H. A. (1937). A direct spectrum of the structure and shift of the Compton line with helium gas as the scatterer. *Physical Review, 52*(5), 419–429.
21. Everett, D. B., Fleming, J. S., Todd, R. W., & Nightingale, J. M. (1977). Gamma-radiation imaging system based on the Compton effect. In *Proceedings of the Institution of Electrical Engineers* (Vol. 124, pp. 995–1000). IET Digital Library.
22. Frieden, B. R., & Soffer, B. H. (2009). De Broglie's wave hypothesis from Fisher information. *Physica A: Statistical Mechanics and its Applications, 388*(7), 1315–1330.

23. Germer, L. H. (1925). The distribution of initial velocities among thermionic electrons. *Physical Review, 25*(6), 795–800.

24. Giancoli, D. C. (2008). *Physics for Scientists and Engineers with Modern Physics* (Vol. 2). Pearson Education.

25. Gil, D., & Solbes, J. (1993). The introduction of modern physics: Overcoming a deformed vision of science. *International Journal of Science Education, 15*(3), 255–260.

26. Glover, G. H. (1982). Compton scatter effects in CT reconstructions. *Medical Physics, 9*(6), 860–867.

27. Hunter, G., & Alexandrescu, C. (2005). Quantification of Einstein's concept of the photon experimentally confirmed by the photoelectric effect. In *Photonic Applications in Nonlinear Optics, Nanophotonics, and Microwave Photonics* (Vol. 5971, pp. 151–158). SPIE.

28. Johnson, C. (2012). Mathematical physics of blackbody radiation. *Icarus Education, 19*, 87–90.

29. Katzer, M., Selig, M., & Knorr, A. (2024). Fermionic versus bosonic thermalization in phonon-driven exciton dynamics: An analytic dimensionality study. *Physical Review B, 109*(15), 155110.

30. Khadka, C. B. (2022). Redefinition of De-Broglie wavelength associated with material particle. *Indian Journal of Advanced Physics, 2*(1), 14–16.

31. Klassen, S. (2011). The photoelectric effect: Reconstructing the story for the physics classroom. *Science & Education, 20*, 719–731.

32. Koc, S. (1957). The quantum efficiency of the photoelectric effect in germanium for the 0.3–2 μm wavelength region. *Czech Journal of Physics, 7*(1), 91–95.

33. Kotur, B. R. (2023). Novel equation for photo-electric effect. *IOSR Journal of Applied Physics (IOSR-JAP), 15*(4), 1–4.

34. Krajewska, K., & Kamiński, J. Z. (2012). Compton process in intense short laser pulses. *Physical Review A: Atomic, Molecular, and Optical Physics, 85*(6), 062102.

35. Kramm, G., & Mölders, N. (2009). Planck's blackbody radiation law: Presentation in different domains and determination of the related dimensional constants. *arXiv preprint arXiv:0901.1863, 9*(1), 542–550.

36. Kuhn, T. S. (1961). The function of measurement in modern physical science. *Isis, 52*(2), 161–193.

37. Liu, B. H., Sun, F. W., Gong, Y. X., Huang, Y. F., Ou, Z. Y., & Guo, G. C. (2008). Demonstration of the three-photon de Broglie wavelength by projection measurement. *Physical Review A: Atomic, Molecular, and Optical Physics, 77*(2), 023815.

38. Livi, R., Pettini, M., Ruffo, S., & Vulpiani, A. (1987). The ultraviolet problem and analytical properties of classical field theories. *Journal of Physics A: Mathematical and General, 20*(3), 577.

39. Matsumoto, T., & Tomita, M. (2010). Modified blackbody radiation spectrum of a selective emitter with application to incandescent light source design. *Optics Express, 18*(10), 192–200.

40. McGovern, J. A., Phillips, D. R., & Grießhammer, H. W. (2013). Compton scattering from the proton in an effective field theory with explicit Delta degrees of freedom. *The European Physical Journal A, 49*, 1–20.

41. Menzel, R., Puhlmann, D., Heuer, A., & Schleich, W. P. (2012). Wave-particle dualism and complementarity unraveled by a different mode. *Proceedings of the National Academy of Sciences, 109*(24), 9314–9319.

42. Millikan, R. A. (1916). Einstein's photoelectric equation and contact electromotive force. *Physical Review, 7*(1), 18.

43. Millikan, R. A., & Winchester, G. (1907). XII. The influence of temperature upon photo-electric effects in a very high vacuum, and the order of photo-electric sensitiveness of the metals. *The London, Edinburgh, and Dublin Philosophical Magazine and Journal of Science, 14*(79), 188–210.

44. Morrison, M. (1986). More on the relationship between technically good and conceptually important experiments: A case study. *The British Journal for the Philosophy of Science, 37*(1), 101–115.

45. Niaz, M., Klassen, S., McMillan, B., & Metz, D. (2010). Reconstruction of the history of the photoelectric effect and its implications for general physics textbooks. *Science Education, 94*(5), 903–931.

46. Olpin, A. R. (1931). An interpretation of the selective photo-electric effect from two-component cathodes. *Physical Review, 38*(9), 1745.

47. Pais, A. (1979). Einstein and the quantum theory. *Reviews of Modern Physics, 51*(4), 863.

48. Pialucha, T., Guyott, C. C. H., & Cawley, P. (1989). Amplitude spectrum method for the measurement of phase velocity. *Ultrasonics, 27*(5), 270–279.

49. Polzin, K. L., & Lvov, Y. V. (2017). An oceanic ultra-violet catastrophe, wave-particle duality and a strongly nonlinear concept for geophysical turbulence. *Fluids, 2*(3), 36.

50. Poole, H. H., & Atkins, W. R. G. (1929). Photo-electric measurements of submarine illumination throughout the year. *Journal of the Marine Biological Association of the United Kingdom, 16*(1), 297–324.

51. Prange, R. E. (1958). Dispersion relations for Compton scattering. *Physical Review, 110*(1), 240.

52. Prokhorov, A. V., Hanssen, L. M., & Mekhontsev, S. N. (2009). Calculation of the radiation characteristics of blackbody radiation sources. *Experimental Methods in the Physical Sciences, 42*, 181–240.

53. Radyushkin, A. V., & Weiss, C. (2001). Deeply virtual Compton scattering amplitude at the tree level: Transversality, twist-3, and factorization. *Physical Review D, 63*(11), 114012.

54. Reggiani, L., & Alfinito, E. (2022). Revisiting the Boltzmann derivation of the Stefan law. *Fluctuation and Noise Letters, 21*(05), 2230001.

55. Robitaille, P. M. (2008). Blackbody radiation and the carbon particle. *Progress in Physics, 3*, 36.

56. Ruoff, A. L. (1967). Linear shock-velocity-particle-velocity relationship. *Journal of Applied Physics, 38*(13), 4976–4980.
57. Stenner, M. D., Gauthier, D. J., & Neifeld, M. A. (2003). The speed of information in a 'fast-light' optical medium. *Nature, 425*(6959), 695–698.
58. Sunyaev, R. A., & Khatri, R. (2013). Unavoidable CMB spectral features and blackbody photosphere of our Universe. *International Journal of Modern Physics D, 22*(07), 1330014.
59. Theimer, O. (1971). Derivation of the blackbody radiation spectrum by classical statistical mechanics. *Physical Review D, 4*(6), 1597.
60. Van Vleck, J. H. (1924). The absorption of radiation by multiply periodic orbits, and its relation to the correspondence principle and the Rayleigh-Jeans law. Part I. Some extensions of the correspondence principle. *Physical Review, 24*(4), 330.
61. Varró, S. (2006). A study on black-body radiation: Classical and binary photons. *Acta Physica Hungarica Series B: Quantum Electronics, 26*, 365–389.
62. Wang, J., Casati, G., & Benenti, G. (2022). Classical physics and blackbody radiation. *Physical Review Letters, 128*(13), 134101.
63. Wear, K. A. (2000). Measurements of phase velocity and group velocity in human calcaneus. *Ultrasound in Medicine & Biology, 26*(4), 641–646.
64. Wein, L. M. (1990). Scheduling networks of queues: Heavy traffic analysis of a two-station network with controllable inputs. *Operations Research, 38*(6), 1065–1078.
65. Wein, O., & Ulbrecht, J. (1972). Residence time distribution in laminar flow systems. II. Non-Newtonian tubular flow. *Collection of Czechoslovak Chemical Communications, 37*(10), 3240–3259.
66. Weinberger, O. (1991). Law, institution and legal politics: Fundamental problems of legal theory and social philosophy. *Springer Science & Business Media*. (Vol. 14, pp. 25–28).
67. Whitham, G. B. (1960). A note on group velocity. *Journal of Fluid Mechanics, 9*(3), 347–352.
68. Wolf, S., Lisin, V., Kondratiev, R., Massone, A. M., Galler, G., Ahrens, J., & Wissmann, F. (2001). Compton scattering by the proton using a large-acceptance arrangement. *The European Physical Journal A: Hadrons and Nuclei, 12*(2), 231–252.
69. Wong, D., Lee, P., Shenghan, G., Xuezhou, W., Qi, H. Y., & Kit, F. S. (2011). The photoelectric effect: Experimental confirmation concerning a widespread misconception in the theory. *European Journal of Physics, 32*(4), 1059.

CHAPTER

2 Quantum Physics

LEARNING OBJECTIVES

After studying this chapter, you will be able to:

1. Understand the statement and implications of Heisenberg's Uncertainty principle.
2. Apply Heisenberg's Uncertainty Principle to position and momentum, and to energy and time, with illustrative examples.
3. Understand the concept of wave-particle duality and its significance in quantum physics.
4. Grasp the properties of wave functions and the fundamental postulates of quantum mechanics.
5. Explain the concept of probability density and the process of normalizing wave functions.
6. Set up and derive the one-dimensional, time-independent Schrödinger wave equation, and solve it for specific examples.
7. Explore applications of the Schrödinger wave equation such as bound states, tunneling phenomena, quantum confinement, and scattering problems.
8. Delve into advanced topics including quantum harmonic oscillators, quantum tunneling in condensed matter physics, cosmology, quantum entanglement, and technological applications of quantum mechanics.

KEY TERMS FROM THIS CHAPTER

Eigenfunctions

Heisenberg's uncertainty principle

Probability density

Quantum tunneling

Eigenvalues

Normalization

Quantum entanglement

Schrödinger wave equation

UNIT INTRODUCTION

Quantum mechanics, a fundamental theory in physics, provides a comprehensive understanding of the intricate dynamics exhibited by energy and matter at minuscule levels, including atoms and subatomic elements. This chapter offers a through exploration of the fundamental ideas, principles, and practical applications of quantum mechanics within engineering physics.

In the first decades of the 20th century, a significant shift occurred in the field of physics, as the foundations for quantum mechanics emerged, challenging the long-standing principles of classical mechanics developed by Newton. Traditional mechanics, although effective in describing large-scale systems, falls short when it comes to understanding phenomena at the atomic and subatomic levels. Quantum mechanics address these limitations by introducing fascinating concepts such as wave-particle duality, quantization, and probability.

An essential principle in this field is the Uncertainty Principle, proposed by Werner Heisenberg in 1927. It states that certain pairs of physical properties cannot be measured at once with arbitrary precision, such as position and momentum. This principle is represented mathematically as follows (Piron, 1976):

$$\Delta x \times \Delta p \geq \frac{h}{2}$$

In this context, Δx represents the uncertainty within position, Δp represents the uncertainty within momentum, and h symbolizes the reduced Planck constant. The principle discussed here has far-reaching implications when it comes to comprehending the behavior of the particles at the quantum level.

The wave function, represented by (ψ), is a crucial concept within the field of quantum mechanics. It offers a probabilistic explanation of the squantum status of a system. The squared modulus for the wave function, $|\psi|^2$, provides a representation of a probability density for locating a particle within different regions of space. For the function of waves to possess any physical significance, it must be properly normalized. This ensures that the overall probability of locating the particle within all of space is precisely one (Zeilinger, 1999).

$$\int_{-\infty}^{\infty} |\psi(\pm)|^2 \quad = 1$$

The Schrödinger wave formula serves as the cornerstone for understanding the intricate dynamics of quantum systems. The time-independent form is expressed as:

In this context, the (H) Hamiltonian operator plays a crucial role as it represents the total energy of the system. The energy eigenvalue, denoted by E, and the wave function, represented by (ψ). The Schrödinger equation is a powerful tool for understanding how energy levels and wave functions of a given system behave (Leonhardt, 2003).

The particle in a box problem is a classic example that showcases the practicality of the Schrödinger equation. It explores the scenario of a particle being confined within a one-dimensional potential well, with infinitely high walls. The problem at hand provides a solution that reveals discrete energy levels and corresponding wave functions, which effectively describe the behavior of the particle confined within the box. The energy levels can be determined by:

$$E = \frac{n^2 \hbar^2 \pi^2}{2mL^2}$$

It is important to note that (n) must be a positive integer, (m) represents the mass of a particle, and (L) corresponds to the length of the box.

In this chapter, one will thoroughly examine these concepts, beginning with Heisenberg's Uncertainty Principle before moving on to the features and importance of the wave function. Next, one will explore the Schrödinger wave equation and its practical applications, offering a comprehensive grasp of quantum mechanics within the realm of engineering physics (Busch et al., 2009).

2.1. HEISENBERG'S UNCERTAINTY PRINCIPLE

Here is a comprehensive breakdown of Heisenberg's Uncertainty principle.

2.1.1. Statement of Heisenberg's Uncertainty Principle

The Uncertainty Principle, proposed by Werner Heisenberg in 1927, establishes a fundamental limit in quantum mechanics on the accuracy of simultaneously determining specific combinations of a particle's physical characteristics (Barukčić, 2016). The relationship between the uncertainties in measuring the positioning of a particle (Δx) and momentum (Δp) is governed by a fundamental principle derived from the reduced Planck constant $(\hbar)$.

According to this principle, there is a specific threshold below which the product of these uncertainties is unable to fall (Ozawa, 2015).

$$\Delta x \times \Delta p \geq \frac{h}{2}$$

This inequality demonstrates that the precision of measuring a particle's position is inversely proportional to the uncertainty in its momentum, and vice versa. The inherent limit mentioned here is not a result of technological constraints but rather arises from the wave-like behavior exhibited by particles at the quantum scale (Maor, 2009).

When it comes to practical applications, let's take the example of an electron. If one tries to pinpoint its position with absolute accuracy, the act of measuring it disrupts its momentum, which renders it impossible to simultaneously determine both properties with complete precision. This principle highlights the fundamental idea of wave-particle duality, in which particles demonstrate properties of both particles and waves (Lewis & Mayer, 1929).

The Uncertainty Principle formulated by Heisenberg has far-reaching consequences for quantum networks and forms the basis for the probabilistic aspects of quantum mechanics. It presents a fascinating challenge to the traditional notion of determinism, revealing that even at a microscopic scale, particles do not have fixed trajectories or properties that can be reliably determined. Rather, the behavior of particles are determined by probabilities given by the wave function (Werner & Farrelly, 2019). This principle is important in a wide range of scientific and technological advancements. It is particularly significant in the field of quantum computing, where accurate oversight and manipulation of quantum states are vital. It has a significant impact on various fields, including spectroscopy, where the limit of resolution within measuring atomic and molecular spectra is determined by the uncertainty principle (Busch et al., 2007).

2.1.2. Heisenberg's Uncertainty Principle Applied to Position and Momentum

The lack of certainty within the position (Δx) can be calculated by the standard deviation for the position measures (Hilgevoord, 1996).

$$\Delta x = \sqrt{\{(x - x)^2\}}$$

Likewise, the definition of uncertainty in momentum \(Δp) seems

$$\Delta p = \sqrt{\{(p - p)^2\}}$$

The wave function $\psi(x)$ is a crucial mathematical tool in quantum mechanics that allows us to describe the momentum and place of a particle. There is a trade-off between knowing the position of a particle with high precision and knowing its momentum accurately. The more precisely one knows the position (i.e., smaller $\Delta x)$, the less precisely one can determine its momentum (i.e., larger $\Delta p)$, and vice versa (Franke-Arnold et al., 2004). The principle in question emerges because of the wave-like characteristics exhibited by particles in the realm of quantum mechanics. The relationship between a particle's position and momentum is that both are conjugate variables, which implies that it gets connected through a Fourier transform. Therefore, a wave packet that is confined to a specific area in space (small $\Delta x)$ matches a wave packet that is spread out in momentum space (large $\Delta p)$, and vice versa (Wagner, 2021).

Remember

Heisenberg's Uncertainty Principle states that the more precisely the position of a particle is determined, the less precisely its momentum can be known, and vice versa.

2.1.3. Heisenberg's Uncertainty Principle applied to Energy and Time

The relationship between uncertainty in energy (ΔE) and the lifetime of a quantum state is well-established, just as the uncertainty in time evaluation is represented by (Δt). According to this principle, there is a relationship between the precision of energy determination in a quantum state along the uncertainty within the time for measurement. Specifically, when the energy determination is more precise (i.e., smaller ΔE), the uncertainty within the measurement time is greater (i.e., larger Δt), and vice versa (Kijowski, 1974).

$E=\hbar\omega$ The uncertainty principle involving energy and time is especially significant in the examination of unstable particles and quantum mechanical processes. In these cases, the lifetime (τ) of a particle is connected to its energy uncertainty (Dutta Majumder & Dutta, 2007).

$\Delta E\,\tau \approx \hbar$ This relationship demonstrates that a particle with a clearly defined energy has a limited duration of existence, and vice versa. Quantum field theory delves deeper into these ideas, incorporating the concept of uncertainty to explain the dynamics of particle-field interactions over time (Agravat, 2013). The uncertainty principle has far-reaching applications in various fields, including quantum optics, spectroscopy, and particle physics. It highlights the inherent limits to the accuracy of measurements and the probabilistic behavior of quantum systems, which are crucial for comprehending theory and implementing practical applications in contemporary physics (Hilgevoord, 1998).

2.1.4. Illustration: Heisenberg Microscope

The Heisenberg Microscope thought experiment demonstrates the practical consequences of Heisenberg's Uncertainty Principle when attempting to measure the position or momentum of subatomic particles, particularly electrons, using photons (Hadzidaki, 2008) (Figure 2.1).

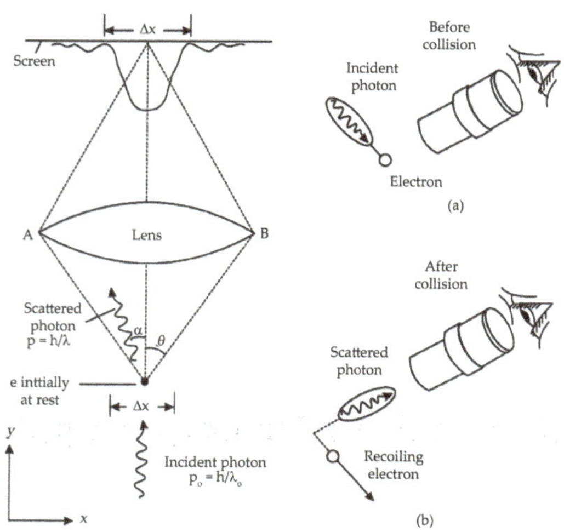

Figure 2.1. *Experimental set-up to measure the position and linear momentum.*

Source: J.S. Aithal, Creative Commons License.

Consider a theoretical situation where a microscope uses photons, which are particles of light, to ascertain the precise location of an electron. According to classical optics, enhancing spatial resolution would require the use of photons with shorter wavelengths (greater energy), allowing for more precise position measurements. Yet, in the realm of quantum mechanics, the interplay between photons and electrons gives rise to an inherent constraint (Habibi et al., 2016).

The uncertainty principle states that there exists an inherent compromise between the accuracy of location (Δx) and momentum (Δp) measurements.

When trying to decrease (Δx) by using shorter wavelengths, the corresponding increase in (Δp) results in a higher level of uncertainty regarding the electron's momentum. This relationship arises as a direct result of a fundamental principle that emphasizes the inherent limitations in measuring both the position and momentum of particles simultaneously (Prasad & Roychoudhuri, 2011).

In addition, the Heisenberg microscope demonstrates the uncertain nature of quantum measurements. Instead of producing predictable results, the interaction between photons and electrons leads to uncertain distributions, highlighting the wave-like characteristics of particles in the realm of quantum mechanics.

Practically, the far-reaching consequences of the Heisenberg Microscope can be observed in various fields like electron microscopy and nanotechnology, where accurate measurements of particle properties are crucial. Understanding and navigating through these mind-bending quantum phenomena is crucial for correctly interpreting experimental findings and pushing forward with technologies rooted in quantum principles (Bastardis et al., 2007).

2.2. APPLICATIONS OF UNCERTAINTY PRINCIPLE

Based on an EEG signal, this study explores the use of the Uncertainty Principle within the field of poor signal detection, specifically denoising. The concept explores the precision of measurement in the realm of quantum physics, and it holds significant consequences in various fields of modern science and engineering. One of these consequences was previously discussed. It presents exciting opportunities for exploring new research paths in denoising and identifying weak digital signals, which could have practical applications in technology solutions (Krenn & Zeilinger, 2020).

2.2.1. Wave-Particle Duality

Wave-particle duality is a fundamental concept in quantum physics, asserting that particles and quantum phenomena can be described as either waves or particles. Challenges arise when attempting to describe the behavior of objects at the quantum scale using classical language, such as terms like "particle" and "wave" (Rempe, 2000).

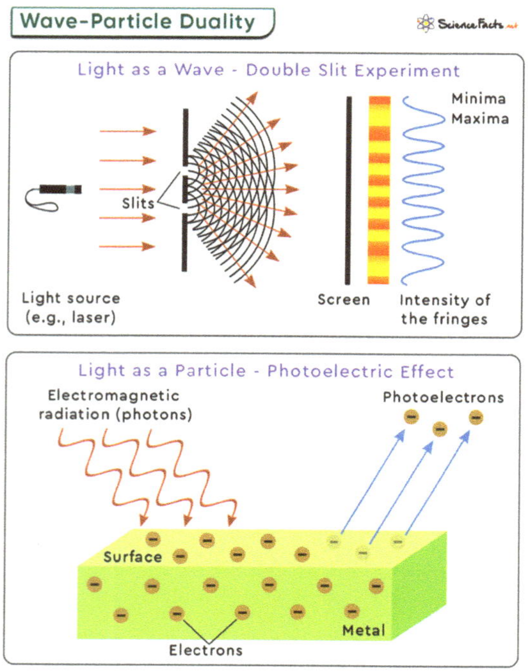

Figure 2.2. *Illustration of the wave particle duality.*

Source: Science Facts, Creative Commons License.

According to the research of Max Planck, Niels Bohr, Albert Einstein, Louis de Broglie, Erwin Schrödinger, Arthur Compton, and various other scientists, it is currently believed that particles and waves are interconnected in the realm of scientific theory. This phenomenon has been noted not only in elementary particles but also in composite particles like atoms and molecules. Wave characteristics often go undetected in larger particles because of their shorter wavelengths (Chang, 2021) (Figure 2.2).

Did you know?

Wave-particle duality illustrates that particles, such as electrons, exhibit both wave-like and particle-like behaviors depending on the experimental setup.

2.2.2. Velocity

The understanding of velocity likely aligns with the scientific definition. One already knows that a substantial change in position over a brief duration corresponds to a high speed, and that speed is quantified in terms of distance divided by time units such as miles per hour, or kilometers per hour.

The average velocity can be calculated by dividing the displacement by the duration of the travel (Bunge, 2003).

$$v_{avg} = \frac{\Delta x}{\Delta t} = \frac{x_f - x_0}{t_f - t_0}$$

Quantum mechanics introduces a fascinating concept called the Uncertainty Principle, which sets a fundamental limit on how precisely one can predict specific features of a particle, like its position and momentum, based on initial conditions. This principle is expressed through various mathematical inequalities (Shirokov, 1978).

2.3. WAVE FUNCTION

The wave function in quantum physics provides a mathematical representation of a particle's quantum state, incorporating factors such as time, momentum, position, and spin. The wave function is typically represented by the Greek letter psi, Ψ (Raymer, 1997) (Figure 2.3).

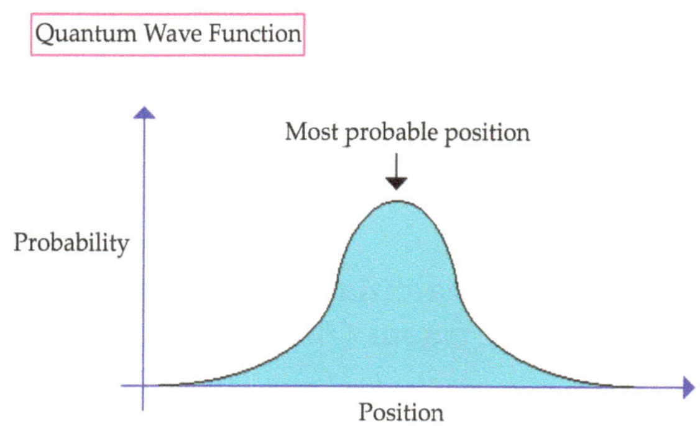

Figure 2.3. *Visualization of quantum wave function.*

Source: University of Oregon, Creative Commons License.

By utilizing a wave equation, it becomes possible to elucidate the likelihood of locating an electron in a matter wave. One way to arrive at this solution is by incorporating an imaginary number, which, when squared, yields a real number. This ultimately determines the position of the electron. The wave function was first introduced in 1925 through the utilization of the Schrodinger equation (Eckhardt et al., 1995).

2.3.1. Properties of Wave Function

Here are some characteristics of the wave function:

1. All the measurable information regarding the particle is readily accessible.
2. ⊠ should exhibit continuity and have a unique value.
3. With the equation of Schrodinger, energy equations become much simpler.
4. The establishment of probability distribution within three dimensions is achieved through the utilization of the wave function.
5. The probability of detecting a particle, assuming its existence, is 1 (Christiansen, 2007).

2.3.2. Postulates of Quantum Mechanics

The subsequent statements outline the fundamental principles of quantum mechanics:

1. By utilizing the time-dependent Schrodinger equation, one can determine the temporal progression of the wave function.
2. Understanding the system becomes easier when considering a particle within a conservative field of force system and utilizing the wave function.
3. A linear set of self-sufficient functions is created by using a set of eigenfunctions from a specific operator.
4. The operator Q, which is associated with the physically quantified property q, is Hermitian.
5. By integrating what is expected about the system's wave function, one can determine the anticipation value of a certain property.
6. For each measurable physical quantity q, an operator Q exists that acts on the wave function linked to a specific value of that quantity, resulting in a wave function multiplied by that value (Marshman & Singh, 2019).

2.3.3. Using the Wavefunction

Insights into the physical nature of a wavefunction Ψ(x,t) can be gained from observing the two-slit diffraction of monochromatic light, which exhibits characteristics similar to electromagnetic waves. The wave function for a light wave is represented by E(x,t), and its level of energy is determined by the square of the electric field strength, denoted as$|E|^2$ (Sebens, 2019). The energy of a single photon is determined solely by the frequency of light, $\epsilon_{photon}=hf$, so $|E|^2$. Therefore, the magnitude of the energy is directly proportional to the number of photons present. When light waves from source S1 interact with light waves from source S2 on the viewing screen, a fascinating interference pattern emerges. The appearance of bright fringes is a result of constructive interference of light waves, while the presence of dark fringes is caused by destructive interference of light waves (Ringbauer et al., 2015).

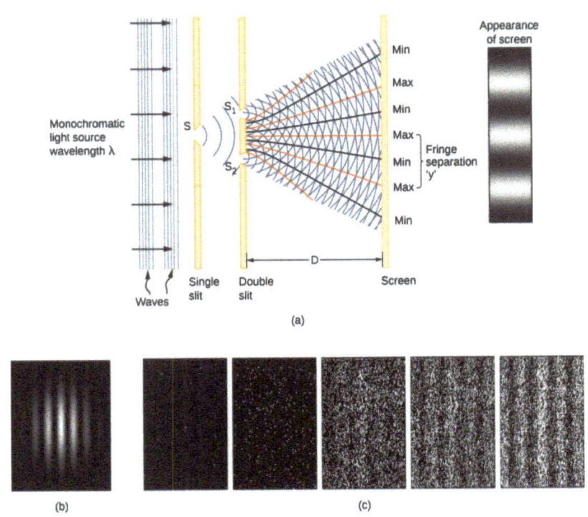

Figure 2.4. *Two-slit interference of monochromatic light. (a) Schematic of two-slit interference; (b) light interference pattern; (c) interference pattern built up gradually under low-intensity light (left to right).*

Source: Openstax, Creative Commons License.

Imagine a screen that starts without any exposure to light. When the screen is subjected to extremely faint light, an interference pattern slowly emerges from left to right (Figure 2.4c). The screen displays individual photons as dots. The density of these dots will be higher at locations where the interference pattern will ultimately become strongest (Gros, 1989). In short, the likelihood of a single photon hitting a specific spot on the screen is directly related to the square for the entire electric field $|E|^2$ at that point. Under specific circumstances, matter particles like electrons can exhibit the same interference pattern (Tannous et al., 2008).

The square of the matter-wave $|\Psi|^2$ in one dimension can be interpreted similarly to the square of the electric field $|E|^2$. The probability of locating a particle at a specific position over time per unit length is provided, commonly referred to as the probability density. The likelihood (P) of locating a particle within a small range $(x, x+dx)$ at a given time t is thus

$$P(x, x+dx) = |\Psi(x,t)|^2\, dx.$$

The wavefunction's probabilistic interpreting is commonly referred to as the Born interpretation. Figure 2.5 provides instances of wavefunctions along with their squares at a specific time t (Lundeen et al., 2011).

Figure 2.5 showcases various examples of wave functions along with their corresponding squared wave functions. If the wavefunction changes gradually over the period, Δx, the likelihood of finding a particle in that interval is roughly

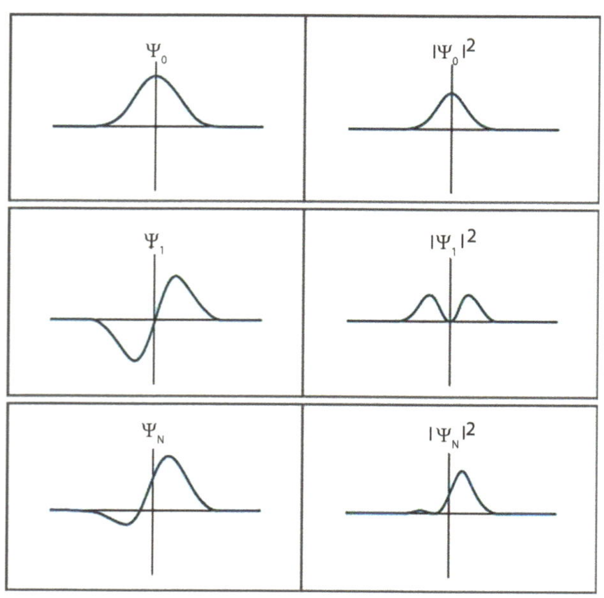

Figure 2.5. *Various examples of wave functions.*

Source: Samuel J. Ling, Creative Commons License.

$$P(x, x+\Delta x) \approx |\Psi(x,t)|^2 \delta x.$$

It is important to observe that when the wave function is squared, the resulting probability is always positive. (This is like raising the strength of an electric field to the power of two, resulting in an increased level of intensity regardless of the field's initial polarity.) However, if the wave function is not shifting gradually, one needs to integrate (Ord, 2009):

$$P(x, x+\Delta x) = \int x+\Delta x x |\Psi(x,t)|^2 \, dx.$$

The probability can be determined by calculating the region within the function $|\Psi(x,t)|^2$ within the range of x and $x+\Delta x$ The likelihood of locating the particle "somewhere" was

$$P(-\infty, +\infty) = \int \infty - \infty |\Psi(x,t)|^2 \, dx = 1.$$

When dealing with particles in different dimensions, the integration process varies. In two dimensions, the integration is performed over an area and involves a double integral. On the other hand, in three dimensions, an integration is carried out over a volume and requires a triple integral. For now, let's focus on the basic one-dimensional scenario (Weinacht et al., 1999).

2.4. PROBABILITY DENSITY AND NORMALIZATION OF WAVE FUNCTION

The waveform $\psi(x,t)$ is a crucial element in understanding the quantum state of a particle in the realm of quantum mechanics (Fröman, 1978). The wave function encompasses all the necessary details of a system and is employed to compute the probability density, that determines the chances of locating a particle within a specific area in space. For a wave function to have physical significance, it needs to be properly normalized (Flatté et al., 1994).

2.4.1. Probability Density

The probability density function, or $P(x,t)$ is obtained by deriving it by using the wave function $\psi(x,t)$ which is precisely defined as

$$P(x,t) = |\psi(x,t)|^2$$

In this context, $P(x,t)$ denotes the probability density, a function that is both real and non-negative, serving to indicate the likelihood of finding a particle at the position (x) at a given time (t).

In the context of one dimension, if the function $\psi(x,t)$ has complex values, the probability density can be determined as follows (Silverman, 1982):

$$P(x,t) = \psi^*(x,t)\,\psi(x,t)$$

The expression $\psi^*(x,t)$ represents the complex conjugation of ψ(x,t).

2.4.2. Normalization of Wave Function

Normalization requirement guarantees that the overall probability of locating the particle throughout all space is precisely 1. Mathematically, this is represented or formulated as:

$$\int_{-\infty}^{\infty} |\psi(x,t)|^2 \, dx = 1$$

For the wave function to be normalized, it must meet a specific condition. If the wave function is not properly normalized, a simple solution is to divide by a square root for the total probability (Zhou et al., 2021).

Let's look at the wave function $\psi(x,t)$. To normalize it, the first step is to calculate the integral.

$$\int_{-\infty}^{\infty} |\psi(x,t)|^2 \, dx = C$$

Where, (C) represents a constant. When normalizing $\psi(x,t)$, it is divided by $\sqrt{C}$ for proper adjustment.

$$\psi_{norm}(x,t) = \frac{\psi(x,t)}{\sqrt{C}}$$

The standardized wave function of $\psi_{norm}(x,t)$ (x,t) fulfills the normalization requirement (Kim et al., 2006).

$$\int_{-\infty}^{\infty} \psi_{norm}(x,t)^2 \, dx = 1$$

2.4.2.1. Normalization of a Gaussian Wave Packet

Let's look at the Gaussian wave packet that is given by:

$$\psi(x,0) = A e^{-\frac{x^2}{2\sigma^2}}$$

where (A) represents the amplitude or (σ) denotes the standard deviation

To regulate the wave function, one need to find the normalizing constant (A) so that

$$\int_{-\infty}^{\infty} |\psi(x,0)|^2 \, dx = 1$$

Tip

When normalizing a wave function in quantum mechanics, ensure that the integral of the probability density over all space equals one, indicating that there is certainty of finding the particle somewhere in space.

Placing $\psi(x,0)$:

$$\int_{-\infty}^{\infty} \left| Ae^{-\frac{x^2}{2\sigma^2}} \right| dx = 1$$

$$A^2 \int_{-\infty}^{\infty} \left| e^{-\frac{x^2}{\sigma^2}} \right| dx = 1$$

Utilizing the result of the Gaussian integral:

$$\int_{-\infty}^{\infty} \left| e^{-\frac{x^2}{\sigma^2}} \right| dx = \sqrt{\pi\sigma}$$

One get:

$$A^2 \sqrt{\pi\sigma} = 1$$

Resolving for \(A):

$$A = \frac{1}{\left(\pi\sigma^2\right)^{\frac{1}{4}}}$$

Therefore, the standardized Gaussian wave function seems (Pattanayak et al., 1994):

$$\psi(x,0) = \frac{1}{\left(\pi\sigma^2\right)^{\frac{1}{4}}} e^{-\frac{x^2}{\sigma^2}}$$

2.4.2.2. Normalization in Three Dimensions

The normalizing criterion for the three-dimensional wave function called F $\psi(\vec{r},t)$ is expanded to include.

$$\int_{-\infty}^{\infty}\int_{-\infty}^{\infty}\int_{-\infty}^{\infty} \left|\psi(\vec{r},t)\right|^2 d^3r = 1$$

Such as $d^{3r} = dx, dy, dz$.

Using a spherical coordinates system, the normalizing condition is expressed as:

$$\int_0^{\infty}\int_0^{\pi}\int_0^{2\pi} |\psi(r,\theta,\phi,t)|^2 r^2 sin\theta dr\, d\theta\, d\phi = 1$$ e

The probability density function alongside the normalization of the wave function plays a vital role in quantum mechanics. They are essential for ensuring that each wave function accurately represents a physically meaningful and probabilistic state. The probability density function $P(x,t) = |\psi(x,t)|^2$ represents the likelihood of locating a particle at a particular position, while normalization guarantees that the total probability across all space is equal to 1. These principles serve as the foundation for comprehending and applying quantum mechanical theories to practical situations (Park et al., 2022).

2.5. SETTING UP OF A ONE DIMENSIONAL, TIME INDEPENDENT, SCHRÖDINGER WAVE EQUATION

Following is the explanation of one dimensional, time independent, Schrödinger wave equation.

2.5.1. Schrödinger Wave Equation

This Schrödinger wave equation serves as a crucial equation in the realm of quantum mechanics, illustrating the dynamic evolution of a physical system's quantum state. Within the realm of one-dimensional systems with unchanging potentials, our attention is directed towards the time-independent Schrödinger equation (Figure 2.6).

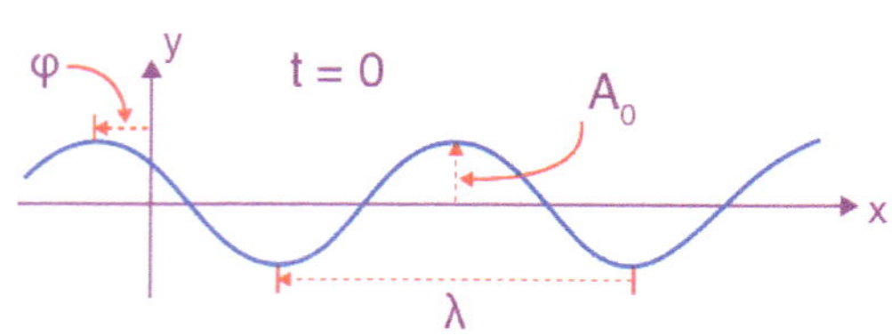

Figure 2.6. *Classical plane wave equation.*

Source: BYJU's, Creative Commons License.

The time-independent Schrödinger equation can be derived by considering situations in which the potential energy $V(x)$ remains constant over time, based on the time-dependent equation.

The equation that describes the behavior of quantum systems as a function of time is known as the time-dependent Schrödinger equation (Zheng, 2007).

$$\psi(x,t) = \phi(x)e^{-iEt/\hbar} \; i\frac{(x,t)}{t} = (-\frac{2}{2m}\frac{2}{x^2} + V(x))(x,t)$$

To derive the time-independent Schrödinger formula, it is postulated that a wave function $\psi(x,t)$ can be decomposed into separate spatial and temporal components.

$$[\psi(x,t) = \phi(x)e^{-iEt/\hbar}]$$

In the given scenario, $\phi(x)$ represents the spatial wave function, while (E) corresponds to the energy of the system.

After substituting this shape into the time-dependent Schrödinger equation and simplifying, the resulting expression is obtained.

$$i\hbar\frac{\partial}{\partial t}\left[\phi(x)e^{-iEt/\hbar}\right] = \left(-\frac{\hbar^2}{2m}\frac{\partial^2}{\partial x^2} + V(x)\right)\phi(x)e^{-iEt/\hbar}$$

Considering the rate of change over time of $e^{-\frac{iEt}{\hbar}}$ is:

$$\frac{\partial}{\partial t}\left(e^{-iEt/\hbar}\right) = -\frac{iE}{\hbar}e^{-iEt/\hbar}$$

On the left-hand side, we observe a transformation:

$$-E\,\phi(x)e^{-\frac{iEt}{\hbar}}$$

By equating the components that do not depend on time, one derives the time-independent Schrödinger equation:

$$-\frac{\hbar^2}{2m}\frac{d^2\phi(x)}{dx^2} + V(x)\phi(x) = E\phi(x$$

The equation characterizes the dynamics of a particle with mass (m) within a potential $V(x)$.

The function $\phi(x)$ provides information regarding the particle's stationary states, while (E) denotes the energy levels associated with these states (Erol, 2010).

2.5.2. Derivation

To obtain the time-independent Schrödinger equation using fundamental principles, one considers the overall structure within the time-dependent Schrödinger equation:

$$i\hbar\frac{\partial\psi(x,t)}{\partial t} = \hat{H}\psi(x,t)$$

A Hamiltonian operator $\hat{(\{H\})}$ encompasses either the kinetic or potential energy components. The Hamiltonian on the particle is one dimension can be defined as

$$\hat{H} = -\frac{\hbar^2}{2m}\frac{\partial^2}{\partial x^2} + V(x)$$

By considering a solution of the form $\psi(x,t)=\phi(x)T(t)$, and substituting this into the equation that depends on time,

$$i\hbar\frac{\partial(\phi(x)T(t))}{\partial t} = \left(-\frac{\hbar^2}{2m}\frac{\partial^2}{\partial x^2} + V(x)\right)\phi(x)T(t)$$

This can be divided into two different equations, each focusing on a certain variable. Regarding the aspect of time, one possesses:

$$i\hbar\frac{dT(t)}{dt} = ET(t)$$

The solution to this differential equation can be obtained by solving it:

$$T(t) = e^{-iEt/\hbar}$$

Regarding the spatial aspect, one possesses:

$$-\frac{\hbar^2}{2m}\frac{d^2\phi(x)}{dx^2} + V(x)\phi(x) = E\phi(x)$$

and the energy eigenvalues. This equation is known as the time-independent Schrödinger equation. It is a second-order differential formula that is used to calculate the wave function $\phi(x)$ along with the energy eigenvalues (E) (Helal & Seadawy, 2009).

To resolve the time-independent Schrödinger equations for particular potentials, we adhere to the following general procedure:

1. **Identify the Potential $V(x)$:** Determine the mathematical expression for a potential energy function V(x) by the given physical problem. Some common examples of potentials include the infinite potential well, harmonic oscillator, finite potential well, and potential barrier.
2. **Substitute and Simplify:** Insert the function V(x) within the Schrödinger equation along with streamlining the resulting differential equation.
3. **Apply Boundary Conditions:** Establish the boundary conditions by considering the physical constraints of the problem. For instance, in the case of an infinite potential well, the wave function needs to be equal to zero at the limits.
4. **Find the Solution:** To the clarified differential equations with the given boundary conditions to determine a wave function $\phi(x)$ and the related energy levels (E) (Cetto et al., 2015).

2.5.3. Example: Infinite Potential Well

Let's examine a particle confined within a potential well that has an infinite width, denoted as (L): (Figure 2.7).

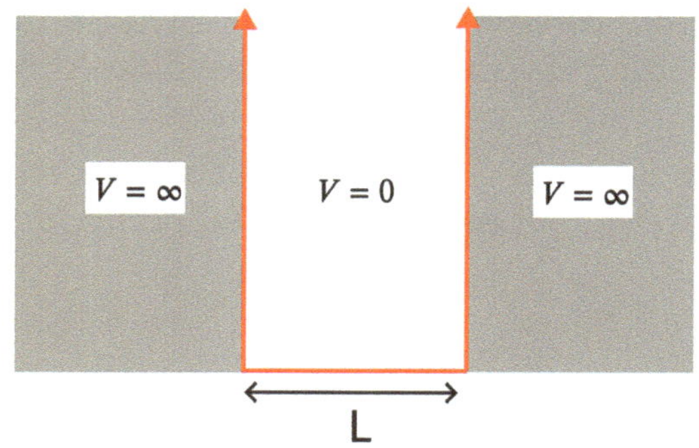

Figure 2.7. *The infinite potential well.*

Source: Dina Zhabinskaya, Creative Commons License.

$$V(x) = \begin{cases} 0 & \text{if} 0 < x < L \\ \infty & \text{if} x \leq 0 \text{or} x \geq L \end{cases}$$

At the point when the potential energy, denoted as $V(x)=0$, is equal to zero, the Schrödinger equation takes on a specific form.

$$-\frac{\hbar^2}{2m}\frac{d^2\phi(x)}{dx^2} = E\phi(x)$$

or

$$-\frac{\hbar^2}{2m}\frac{d^2\phi(x)}{dx^2} = E\phi(x)[\frac{d^2\phi(x)}{dx^2} = -\frac{2mE}{\hbar^2}\phi(x)$$

Simplifying $(k^2 = \frac{2mE}{\hbar^2})$, the equation yields,

$$\frac{d^2\phi(x)}{dx^2} + k^2\phi(x) = 0$$

The solution that applies to all cases is:

$$\phi(x) = A\sin(kx) + B\cos(kx)$$

Implementing border conditions $\phi(0)=0$ and $\phi(L)=0$, one find $(B=0)$ and $(k = \frac{n\pi}{L})$, where (n) represents a positive integer. Yet, the wave functions being normalized are:

$$\phi_n(x) = \sqrt{\frac{2}{L}}\sin\left(\frac{n\pi x}{L}\right)$$

The energy levels that correlate to this are:

$$E_n = \frac{n^2\pi^2\hbar^2}{2mL^2}$$

This derivation demonstrates the solution of the time-independent Schrödinger equation for a particular potential, resulting in discrete energy levels and their corresponding wave functions (Duo & Zhang, 2015).

2.5.4. Eigenvalues and Eigenfunctions

The solutions for the time-independent Schrödinger equation in quantum physics are characterized by their eigenvalues and eigenfunctions. These solutions delineate the potential energy states for a quantum system including the corresponding wave functions that represent these states (Berkdemir et al., 2006).

2.5.4.1. Eigenvalue Problem

The Schrödinger equation, which does not depend on time, might be expressed as an eigenvalue issue:

$\hat{H}\phi(x)=E\phi(x)$ The expression $(\hat{H})$ represents the Hamiltonian operator, while $\phi(x)$ represents a eigenfunction as well as (E) represents the eigenvalue (or energy) (Yang & Zhou, 2021).

2.5.4.2. Physical Interpretation

Within this context:

1. Eigenvalues (E) correspond to the permissible energy states of the system.
2. The eigenfunctions $\phi(x)$ correspond to the probability amplitudes linked to these levels of energy.
3. The eigenfunctions must adhere to precise boundary constraints and meet normalization standards. The eigenfunctions constitute a comprehensive set of ortho functions, allowing for the representation of any arbitrary waveform through a linear blend of these functions (Baily & Finkelstein, 2015).

2.5.5. Example: Particle in a Box

Let's take another example of limitless potential well (Figure 2.8).

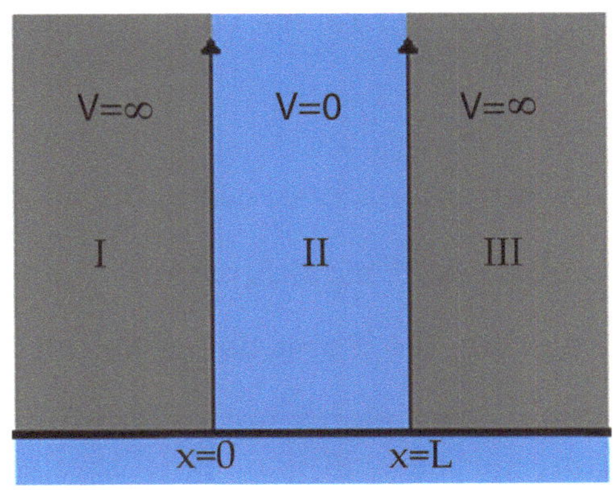

Figure 2.8. *The particle in a box.*

Source: Dan Thomas, Creative Commons License.

The equation that describes this system has become the Schrödinger equation:

$$\frac{d^2\phi(x)}{dx^2} + k^2\phi(x) = 0$$

Given the boundary conditions $\phi(0)=0$ and $\phi(L)=0.$ Here are the answers:

$$\phi_n(x) = \sqrt{\frac{2}{L}}\sin\left(\frac{n\pi x}{L}\right)$$

and the energy levels that correspond to this phenomenon are:

$$E_n = \frac{n^2\pi^2\hbar^2}{2mL^2}$$

Here, (n) seems a positive integer that represents the quantitation of energy levels. Every value of (n) indicates a unique eigenfunction $(E_{n)}$ (x) when eigenvalue (Castro, 2020).

2.5.6. Orthogonality and Normalization

The eigenfunctions of the Hamiltonian operator exhibit orthogonality and can be normalized. Orthogonality refers to a concept in physics that describes the relationship between different elements or states:

$$\int_0^L \phi_m(x)\phi \int_0^L \phi_m(x)\phi_n(x)\,dx = 0 \quad \text{form} \neq n$$

Normalization refers to a process:

$$\int_0^L |\phi_n(x)|^2\,dx = 1$$

The conditions over the infinite potential effectively are automatically satisfied by the standardized eigenfunctions $_{n(x)}$ because of their sine function form.

Exploring the Potential of Quantum Systems

Eigenvalues along with eigenfunctions play a vital role in solving a wide range of quantum mechanical issues. Below are a few illustrations (Alhaidari, 2019):

2.5.6.1. Harmonic Oscillator

In the case of the quantum harmonic oscillator, its potential takes on a specific form (Figure 2.9)

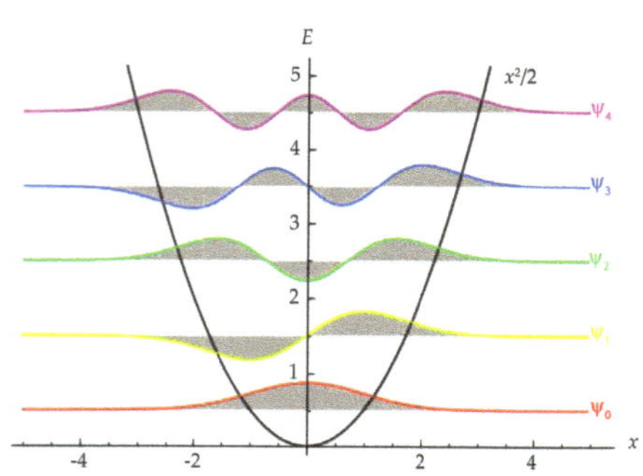

Figure 2.9. *Illustration of simple harmonic oscillator.*

Source: Mohammad Yasir, Creative Commons License.

$(V(x) = \frac{1}{2}m\omega^2x^2)$.

The Schrödinger equation undergoes a transformation.

$$\left(-\frac{\hbar^2}{2m}\frac{d^2}{dx^2} + \frac{1}{2}m\omega^2x^2\right)\phi(x) = E\phi(x)$$

The solutions involve a set of specialized polynomials $H_{n(x)}$ multiplied by a Gaussian factor, and the energy levels exhibit quantization (Kastrup, 2007).

$$E_n = \left(n + \frac{1}{2}\right)\hbar\omega$$

2.5.6.2. Hydrogen Atom

Regarding the hydrogen atom,

$$V(r) = -\frac{e^2}{4\pi\epsilon_0 r}$$

potential it experiences is the Coulomb potential. The formula in question, when expressed in spherical coordinates, is as follows:

$$\left(-\frac{\hbar^2}{2m}\nabla^2 - \frac{e^2}{4\pi\epsilon_0 r}\right)\psi(r,\theta,\phi) = E\psi(r,\theta,\phi)$$

The solutions are determined by spherical harmonics $_{(lm)}(\theta,\phi)$ and radial functions $R_{(nl)}(r)$ thus energy levels are specified by these solutions (Jensen et al., 1991).

$$E_n = -\frac{13.6\,\text{eV}}{n^2}$$

Where, ( ) represents the primary quantum number

Comprehending eigenvalues and eigenfunctions is essential in the field of quantum mechanics. These principles establish a mathematical framework for examining the discrete energy levels and corresponding wave function for quantum systems. By resolving the equation proposed by Schrödinger for various potentials, one gets to understand the behavior and characteristics of particles on the quantum scale (Gaeta & Stroud, 1990).

2.6. APPLICATIONS OF THE SCHRÖDINGER WAVE EQUATION

The Schrödinger wave equation is a fundamental principle in quantum mechanics that dictates the behavior of small-scale systems. The wavefunction, $\Psi(x, t)$, for a quantum system is associated with its total energy (E). This chapter examines multiple uses for the Schrödinger equation, emphasizing its effectiveness in explaining a wide range of phenomena in the quantum domain (Dysthe, 1979).

2.6.1. Bound States in Atoms and Molecules

An essential application of this lies in understanding the configuration of molecules and their atoms. The Schrödinger equation is used to explain the behavior of electrons that are bound to the atomic nucleus due to the electrostatic forces. The equation's potential energy term, $V(x)$, reflects the attractive force. The time-independent Schrödinger equations for this system can be solved to determine the permissible energy levels (eigenvalues) as electrons can occupy. The energy levels mentioned here correlate to the specific electron orbitals that are observed in atoms, including the well-known s, p, d, or f orbitals. The absolute value squared of the wavefunction, denoted as $|\Psi(x,t)|^2$ represents the likelihood density of locating the electron on a certain position within the atom (Stenholm, 1973).

2.6.1.1. Mathematical Formulation

The Schrödinger equation applied to a one-dimensional system, which does not depend on time, might be expressed as:

$$-\frac{\hbar^2}{2m}\frac{\partial^2\Psi(x)}{\partial x^2} + V(x)\Psi(x) = E\Psi(x)$$

Where, $\Psi(x)$ represents the wavefunction; $\hbar$ (h-bar) is the reduced Planck constant ($h/2\pi$); m shows the mass of particle; V(x) shows potential energy function; and E is the total energy for system.

By solving the following equation for different potential energy functions that correspond to various atomic and molecular configurations, one can determine the permissible energy levels and their associated wavefunctions (Kavka et al., 2010).

2.6.2. Tunneling Phenomena

The Schrödinger equation elucidates the intriguing process of quantum tunneling. In the realm of classical physics, a particle must possess sufficient energy to surpass a barrier of potential energy. Nevertheless, the particle's wave-like behavior in quantum physics permits a limited likelihood of the particle tunneling through the barrier, despite lacking adequate classical energy. The particle's wavefunction exhibits a non-zero value across the region that is banned according to classical physics, however its amplitude decays exponentially. The likelihood of successful tunneling is contingent upon the dimensions of the barrier, including its width and height, along with the mass and energy for the particle (Bittner, 2000) (Figure 2.10).

Figure 2.10. *Quantum tunnelling of a lion through a barrier is a quantum effect with no classical analog.*

Source: Ümit Kaya, Creative Commons License.

2.6.3. Quantum Confinement

Quantum confinement arises from applying the Schrödinger equation in systems that have limited spatial dimensions. Let's take, for example, a particle that is restricted to move within a one-dimensional box with walls that have a finite length. The particle is effectively confined within the box by setting the potential energy outside the box to infinity. The quantized energy levels for each confined particle are obtained by solving the Schrödinger equation of this system. These results have significant ramifications in several domains, such as the conduct of electrons within semiconductor devices along with the level of energy for quantum dots (Stucky and Mac Dougall, 1990) (Figure 2.11).

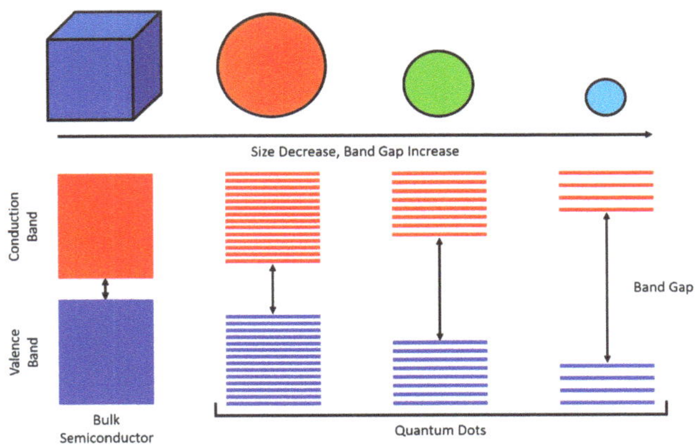

Figure 2.11. *Illustration of quantum confinement effect on the band gap of general semiconductors.*

Source: Matthew W. Brett, Creative Commons License.

2.6.4. Scattering Problems

The Schrödinger equation is often used to examine scattering phenomena, in which a particle interacts with either a potential barrier or a different particle. The particle's incoming and outgoing wavefunctions are determined by solutions to the Schrödinger equation. Through the analysis of the scattered wavefunction, it is possible to ascertain the probability of a particle being scattered at various angles and its distribution for final momentum. This application is essential for understanding diverse phenomena, such as nuclear reactions, electron diffraction, including the dynamics of particles in accelerators.

These examples represent only a small fraction of the extensive range of uses for the Schrödinger equation. The ability of quantum systems to be described probabilistically has greatly transformed our understanding of the microscopic realm, leading to progress in areas such as condensed matter physics, quantum chemistry, and nanotechnology (Faddeev, 1990).

2.7. ADVANCED TOPICS IN QUANTUM MECHANICS

2.7.1. Quantum Harmonic Oscillator

The quantum harmonic oscillator (QHO) is a fundamental system in quantum mechanics. The simplicity of its design, along with its intricate mathematical structure, make it an ideal model for understanding a wide range of quantum phenomena. This chapter examines the theoretical framework of the quantum harmonic oscillator (QHO), investigating its solutions, energy levels, and related wavefunctions (Dekker, 1981).

2.7.1.1. Classical Harmonic Oscillator

Exploring the depths of the quantum realm requires us to take a step back and revisit its classical counterpart, the simple harmonic oscillator (SHO). This system involves a mass (m) connected to a fixed point by a spring with the spring constant (k). When a spring is displaced from its equilibrium position, it generates a force that pushes it back in proportion to the distance it has been displaced (x). This force is described by Hooke's law (F = -kx). leading to a clearly defined equation of motion:

$$m\ddot{x} = -kx$$

where x stands for the second a derivative of x in relation to time (t). By resolving the differential equation, one can observe the well-known oscillation of the mass, which is characterized by an angular frequency (ω):

$$\omega = \sqrt{k/m}$$

The combined energy of the classical Simple Harmonic Oscillator (SHO) is determined by the total of the kinetic energy (E) along with potential energy (V):

$$E = T + V = 1/2m\dot{x}^2 + 1/2kx^2$$

The energy remains unchanged throughout the period of oscillation (Alvarez et al., 1988).

2.7.1.2. Quantization and the Hamiltonian

In the quantum regime, the classical description breaks down. The position and momentum of the particle are no longer exact values, but probabilistic quantities described by wavefunctions. The Schrödinger equation governs the evolution of the wavefunction of the system:

When one enters the realm of quantum physics, the traditional classical description no longer holds true.

The position (x) and the momentum (p) associated with the particles are no longer precise values but rather probabilistic quantities given by wavefunctions. This Schrödinger equation describes how the wave function (Ψ (x, t) of a system evolves over time (Rashkovskiy, 2023).

$$i\hbar\frac{\partial\Psi(x,t)}{\partial t}=\hat{H}\Psi(x,t)$$

Where, The imaginary unit is represented by the symbol I; ℏ (h-bar) is the reduced Planck constant (h/2π); and ˆH The Hamiltonian operator is a representation of the system's entire energy.

In the case of the quantum harmonic oscillator (QHO), the Hamiltonian can be written regarding the position as well as momentum operators (p and x) by utilizing the fundamental commutes connection between them.

$$[\hat{x},\hat{p}]=i\hbar$$

The form of the Hamiltonian for a quantum harmonic oscillator is as follows:

$$\hat{H}=\frac{\hat{p}^2}{2m}+\frac{1}{2}m\omega^2\hat{x}^2$$

This form emphasizes the link among classical along with quantum descriptions. The initial term corresponds to the operator for kinetic energy, while the subsequent term corresponds to the operator for potential energy, like the classical equations (Carosso, 2022).

2.7.1.3. Solving the Schrödinger Equation for the QHO

The time-independent Schrödinger formula can be derived by isolating the temporal aspect from the wavefunction:

$$\hat{H}\Psi(x)=E\Psi(x)$$

By substituted the Hamiltonian with a quantum harmonic oscillator (QHO) and rearranging the equation, a new expression is obtained:

$$-\frac{\hbar^2}{2m}\frac{d^2\Psi(x)}{dx^2}+\frac{1}{2}m\omega^2x^2\Psi(x)=E\Psi(x)$$

This equation is commonly known as the harmonic oscillator equations. Despite its apparent simplicity, finding an exact analytical solution for a potential energy function such as the harmonic potential is not always possible. However, there are powerful mathematical methods that can be utilized to solve it (Mohebbi et al., 2013).

One method involves the use of lowering and raising operators, which are operators that can either create or remove quanta of energy within the system. These operators play a vital role in manipulating wave functions to create energy eigenstates alongside their associated wave functions.

Another method involves utilizing Hermite polynomials, which are a unique set of orthogonal polynomials, to solve the equation. These polynomials possess certain characteristics that make them suitable as foundational functions for calculating the wave functions that make up the quantum harmonic oscillator (Nelson, 1966).

2.7.1.4. Energy Levels and Wavefunctions

The Schrödinger equation of a quantum harmonic oscillator uncovers a series of distinct energy levels, in contrast to the continuous spectrum observed in a classical oscillator. The energy levels are determined by quantization:

$$E_n = \left(n + \frac{1}{2}\right)\hbar\omega$$

when n is non-negative integer (n = 0, 1, 2,...) indicating the quantum number of the state.

The energy levels are evenly spaced by a specific amount of energy, emphasizing the discrete nature of the system. The state called the ground state (n = 0) has the lowest possible energy, unlike in classical physics where it would be zero (Hioe & Montroll, 1975).

The wavefunctions $\psi_n(x)$ in all energy levels may be obtained by solving the Schrödinger equation.

The wavefunctions being referred to here are called harmonic oscillator wavefunctions. They provide information about the probablity of finding the particle at different positions x within the potential well. The expressions are formulated using Hermite polynomials $(H_n(x))$ along with a constant for normalization.

$$_n(x) = N_n H_n(\sqrt{\frac{m\omega}{\hbar}}x)e^{-mx^2/(2\hbar)}$$

The normalization constant guarantees that the likelihood of locating the particle at any point along the real number line is exactly 1. The Hermite polynomials, distinguished by their angle (n), dictate the form and positioning of the wavefunctions (Hofstadter, 1976).

Let's look at the important characteristics of the wavefunctions for the harmonic oscillator:

1. **Ground State (n = 0):** The ground state wave function is a Gaussian function centered at, indicating a high probability of finding the particle near its equilibrium position. However, unlike the classical case where the particle rests perfectly at x = 0, the wavefunction extends slightly outwards due to the Heisenberg uncertainty principle.
2. **Excited States (n > 0):** As the value of a quantum number (n) rises, the wave functions start to display more nodes, which are points where the wave function becomes zero. The number for nodes is equal to one less than the value of n. These nodes represent areas where the likelihood of locating the particle is reduced. The wave functions become more widely dispersed, indicating a greater average kinetic energy over particles in excited states (Rosner et al., 1984).

2.7.1.5. Zero-Point Energy

One important outcome of the quantized energy levels seem to be the presence of zero-point energy (E_0). The minimum energy ($\hbar\omega/2$) of the oscillator in its ground state is the lowest possible energy level it can have.

The existence of a non-zero minimum energy is a result of the particle's wave-like characteristics. According to the uncertainty principle, it is impossible for a particle to have both zero position along with zero momentum at the same time. As a result, there is a constant residual motion linked to the particle in its lowest energy state, which gives rise to what is known as zero-point energy (Wang, 2022).

2.7.1.6. Applications of the Quantum Harmonic Oscillator

The QHO is widely recognized as an essential system in numerous fields of scientific study. The usefulness of this concept arises from the fact that numerous physical systems may be estimated as harmonic oscillators when getting close to their equilibrium positions. Below are a few notable applications:

1. **Molecular Vibrations:** Modeling the oscillations of atoms throughout a molecule may be achieved using the QHO. Through the solution of the Schrödinger equation using the molecule's unique potential energy function, the vibrational pitches and energy levels for the molecule can determine. This knowledge is essential for gaining insight into the molecule's spectroscopic properties (Barbosa et al., 2015).
2. **Crystal Lattices:** The oscillations of atoms inside a crystal lattice could alternatively be regarded as interconnected harmonic oscillators. This enables the examination of phonons, which are quantized waves of sound within materials, and their profound effects on thermal conductivity along with other material characteristics.
3. **Quantum Field Theory:** The QHO serves as the fundamental framework for understanding less complex quantum field theories. By discretizing the displacement field generated by the oscillator, it is possible to create more intricate quantum field concepts that explain the behavior of elementary particles including their interactions (Liu et al., 2004).

The QHO offers not only useful insights in specific physical structures but also serves as a foundation for understanding more complex quant phenomena.

The quantum harmonic oscillator are a model commonly used to investigate the intriguing realm of quantum mechanics. The well-established mathematical framework of this subject enables a comprehensive understanding of quantization, energy levels, and wavefunctions. The QHO not only reveals the paradoxical characteristics of the quantum realm, but also has practical applications in multiple disciplines, establishing it as a fundamental system in our pursuit to understand the tiny realm (Vergel & Villaseñor, 2009).

2.7.2. Quantum Tunneling in Condensed Matter Physics

Quantum Tunneling, an essential aspect of quantum mechanics, serves a crucial role in the field of condensed matter physics. This phenomenon explores the intriguing concept of particles being able to surpass energy barriers that would normally be considered impassable according to classical mechanics. In this section, we explore the complex world of quantum tunneling within condensed matter systems. One delves into its theoretical foundations, examines its various manifestations, and discusses its important technological applications (Cottet et al., 2017).

2.7.2.1. Classical Versus Quantum Tunneling

In classical mechanics, a particle needs enough kinetic energy to surpass a potential energy barrier. When a particle's energy is lower compared to the barrier's height, it is confined to the area that has less potential energy according to classical principles. Nevertheless, in the realm of quantum mechanics, the wave-like behavior of particles brings about the intriguing possibility of particles "tunneling" through barriers, defying classical expectations despite their seemingly inadequate energy.

The tunneling phenomenon occurs due to the wave function Ψ(x, t) that describes the particle. The wavefunction extends to a classically forbidden area throughout the barrier, although its amplitude decays exponentially. The successful tunneling probability is influenced by various factors, including the dimensions of the barrier and the characteristics of the particle involved (Owerre & Paranjape, 2015).

2.7.2.2. Manifestations of Tunneling in Condensed Matter

Various phenomena across condensed matter systems exhibit manifestations of quantum tunneling. Below are a few notable examples:

1. **Superconductivity:** Superconductors exhibit a fascinating phenomenon where electrons join forces to create Cooper pairs, enabling them to effortlessly traverse the energy barrier between the normal along with superconducting states. This phenomenon enables the smooth flow of current without any loss, which is a key feature of super conductors.
2. **Josephson Junctions:** These junctions are composed of two superconductors that are divided by an extremely thin insulating barrier. The Josephson effect is a fascinating phenomenon that arises from the tunneling of Cooper pairs across a barrier. It showcases distinctive current-voltage characteristics and holds great potential for uses in superconducting quantum devices.
3. **Field-Effect Transistors (FETs):** The conductivity channel found in contemporary MOSFETs is modulated by the gate voltage with a tunneling process. Manipulating the flow of current within a transistor can be achieved by regulating the gate voltage, which serves as the foundation for modern electronic devices.

When examining specific materials, the introduction for a magnetic field can modify their possible energy landscape, which in turn affects the likelihood of electron tunneling. Magnetoresistance, a fascinating phenomenon, is utilized in magnetic random-access memory (MRAM) devices (Bâldea et al., 2001).

2.7.2.3. Factors Affecting Tunneling Probability

There are various factors that can affect the likelihood of quantum tunneling within condensed matter systems.

The dimensions of the barrier play a crucial role in determining the probability for tunneling. A narrower along with lower barrier tends to result in a higher probability during tunneling, while a wider and greater barrier reduces this probability. The wavefunction's rot within the barrier results in a decreased likelihood of locating the particle deeper within.

1. **Particle Mass:** Lighter particles exhibit a greater likelihood of tunneling than do heavier particles because of their intrinsic wave-like properties. The de Broglie's wavelength ($\lambda = h/p$) for a particle is inversely related to its mass (where p represents the momentum). A

particle with lower mass exhibits a greater wavelength, facilitating a more extensive "overlap" over the barrier, hence enhancing the possibility of tunneling.

2. **Temperature:** Elevated temperature can augment the likelihood of tunneling by supplying extra thermal energy for the particle, facilitating the surmounting of the barrier (Isakov et al., 2016).

2.7.2.4. Technological Applications of Tunneling

Quantum tunneling is a fundamental concept that supports several technological developments in the field of condensed matter physics.

1. **Scanning Tunneling Microscope (STM):** This device uses the phenomenon of electron tunneling to capture images of surfaces with atomic precision. By placing a sharp probe close to the surface and applying an electrical bias, the resulting tunneling current provides details on the surface's topography.
2. **Tunnel Diodes:** These electronic components rely on the flow of tunneling currents to function. High-frequency oscillators or mixers utilize the negative resistance region observed in their current-voltage characteristics.
3. **Flash Memory:** Data within flash memory devices are stored by the presence or absence of charge carriers that are trapped within potential wells. Tunneling is a vital component in the process of programming and erasing memory cells. It allows for the injection and extraction of charged particles through thin insulation layers.
4. **Quantum Computing:** Researchers are currently investigating tunneling phenomena to advance quantum computing technology. Within specific frameworks, qubits are encoded using the states of superconductive Josephson junctions. This enables the controlled tunneling of Cooper pairs, which in turn allows for the manipulation of qubit states (Amico et al., 2022).

2.7.2.5. Future Directions and Challenges

Exploration into the fascinating realm of quantum tunneling within condensed matter remains at the forefront of technological advancements and the pursuit of scientific knowledge. Here are a few important areas to consider:

1. **Exploration of Novel Materials:** Exploring the realm of materials and their properties has the potential to unlock exciting advancements in various devices, such as superconductors and spintronic devices. By carefully manipulating these materials, one can enhance their tunneling efficiency and pave the way for new functionalities.
2. **Manipulation and Control of Tunneling Processes:** Exploring methods to achieve precise regulation of tunneling events shows great potential for advancing the field of electronic and quantum devices, leading to improved performance and enhanced functionality (Alexeev et al., 2024).
3. **Bridging the Gap Between Classical and Quantum Regimes:** Exploring the transition from the precise realm of quantum physics to the more familiar classical limit poses a

significant challenge. This research has the potential to uncover a more profound comprehension of the underlying principles of quantum mechanics or its relationship with observable phenomena on a larger scale.

Quantum tunneling is an extraordinary phenomenon that bridges the realms of classical and quantum physics. The study of quantum physics in the field of condensed matter physics has been instrumental in driving various technological advancements and holds immense potential for uncovering new and exciting discoveries in the future. As scientists continue to explore the complexities of tunneling process and investigate new materials, the possibilities for groundbreaking applications in the future are immense. Quantum tunneling has the potential to push the limits of condensed matter physics and open up new possibilities for revolutionary technological progress (Gill et al., 2022).

2.7.3. Quantum Mechanics in Cosmology

Cosmology, the study of the universe's creation and early stages, poses a significant challenge to the traditional framework of general relativity in cosmology. The extreme energy density and space-time curvature involved make classical physics inadequate in this realm. However, quantum theory, with its a probabilistic nature and focus on wave-particle duality, becomes essential in achieving an accurate understanding of the cosmos. This chapter explores the intricacies of quantum cosmology, delving into its theoretical underpinnings, the obstacles it encounters, and its potential for shedding light on the mysteries surrounding the development and origins of the universe (Gell-Mann & Hartle, 1996).

2.7.3.1. The Quantum Gravity Conundrum

One of the main challenges in the field of quantum cosmology is finding a way to harmonize the principles of quantum mechanics and gravity. General relativity, a fundamental concept in classical cosmology, explains gravity as the curvature of spacetime due to the existence of energy and mass. Nevertheless, the foundation of quantum mechanics lies in the notion of particles and fields. Describing gravity within the framework of quantum mechanics, where all objects display both wave and particle characteristics, remains a significant obstacle.

Various potential methodologies strive to close this divide (Nomura, 2021):

1. **Loop Quantum Gravity (LQG):** This theory suggests that spacetime is not a smooth continuum, but rather composed of tiny, indivisible units. These units give the universe a granular nature, with a minimum length scale called the Planck length ($lp \approx 1.6 x 106^{35}$ meters). This theory seeks to reconcile general relativity with quantum effects by incorporating quantified geometric loops, with the goal of recovering general the law of relativity at larger scales. The dynamics of LQG are determined by a quantized form for the Einstein field equations. These equations are expressed using loop variables that indicate the geometry of spacetime. The loop variables may be mathematically expressed through operators that act on a Hilbert space. This Hilbert space serves as the basis for creating quantum states of spacetime.
2. **String Theory:** This theory proposes that elementary particles are not infinitesimally small points, but

rather they are vibrating strings which exist in dimensions over the four dimensions of the spacetime as we are familiar with. String theory embraces gravity as an essential force by suggesting that the shut, vibrating strings themselves serve as gravitons, which are hypothetical quantum particles responsible for carrying gravity.String theory aims to consolidate all the basic powers of nature inside a unified framework. Nevertheless, this concept is predominantly theoretical and has obstacles when it comes to scientific confirmation because of the extra spatial dimensions it proposes.

3. **Quantum Field Theory in Curved Spacetime (QFT-CS):** This technique discretizes matter fields into the curved spacetime given by general relativity. A metric tensor $(g_{\mu\nu})$, representing the spatial and temporal structure of spacetime, is transformed into an operator within the framework of quantum field theory. This transformation results in altered versions of the well-known field equations employed in quantum mechanics.

QFT-CS has demonstrated potential in comprehending specific facets of the earlier universe, like the inflationary epoch. It enables the exploration of particle development and annihilation phenomena in the presence of powerful gravitational fields, offering valuable insights into the behavior of matter in extreme circumstances.

These different approaches, including causal dynamical triangulations and asymptotic safety, are part of the ongoing efforts to establish a coherent theory of gravity at the quantum level. Currently, there is no widely agreed-upon solution, and ongoing research is exploring different approaches to bring together these seemingly unrelated frameworks (Gopakumar, 2015).

2.7.3.2. The Elusive Wavefunction of the Universe

An intriguing hypothesis in quantum cosmology involves the concept of a wavefunction of the cosmos, denoted as Ψ(u). The wave function would contain comprehensive information about the entire universe, encompassing its initial conditions, its temporal evolution (expressed from a variable u), and all possible configurations. Nevertheless, there is ongoing debate around the structure and meaning of this wavefunction.

The Schrödinger equation, a fundamental principle of non-relativistic quantum physics, establishes a connection between the wavefunction and all the energy (E) of a system (Höhn, 2019):

$$i\hbar\frac{\partial\Psi(t)}{\partial t} = \hat{H}\Psi(t)$$

However, the primary challenge in quantum cosmology is developing an appropriate Hamiltonian that integrates gravity and accurately characterizes the dynamics of the cosmos. Different approaches to quantum gravity can result in diverse configurations for the Hamiltonian, and therefore, the wave function.

Certain interpretations based on the wave function propose that the universe can exist as a superposition of various states, each represented by distinct wave functions, before experiencing a phenomenon known as decoherence.

The process of decoherence is an intricate event that can cause the wave function to collapse into a single classical state, effectively selecting one of the many configurations within the superposition (Chen, 2019).

Some propose a dynamic multiverse hypothesis where the wavefunction branches out into several potentialities, leading to a multitude of worlds with unique attributes. The Many-Worlds Interpretation, which suggests the existence of an infinite number of universes, has significant philosophical and physical challenges, primarily because most of these realms are beyond our ability to observe (Biswas et al., 1997).

2.7.3.3. Applications of Quantum Mechanics in Cosmology

Although a complete theory on quantum gravity has not yet been developed, quantum mechanics provides useful insights into many cosmic events:

1. **The Big Bang Singularity:** In classical general relativity, a singularity is predicted at the inception of the universe, as the density or curvature of spacetime reaches an infinite value. This singularity presents a substantial obstacle since our current physical laws become invalid under such extreme conditions. Quantum physics suggest that a theory for quantum gravity might potentially address this singularity by explaining the world at densities corresponding to the Planck scale. Loop quantum gravity posits that the discrete structure of spacetime hinders the presence of absolute infinities such as singularities.
2. **Inflation:** The inflated epoch is a theoretical phase of extremely rapid along with exponential expansion believed to have taken place right after the Big Bang. This period is thought to have evened out the universe and prepared the way for the development of large-scale structures. Theorized fluctuations throughout inflation are believed that it played a crucial role. The fluctuations that occurred due to the uncertainty principle within quantum mechanics might have been amplified during inflation and ultimately gave rise to the irregularities that led to the creation of clusters of galaxies for galaxies. Quantum field theory for curved spacetime is frequently utilized in examining inflationary models to analyze the dynamics of matter fields throughout fast expansion. The variations in these quantum fields are thought to be the origin of the minuscule temperature differences found in the Cosmic Microwave Background (CMB) radiation.
3. **Hawking Radiation:** General relativity predicts that black holes produce radiation because of quantum events occurring near the event horizon. Hawking radiation is a phenomenon that occurs due to the continuous creation and annihilation of virtual particle-antiparticle pairs near the event horizon.

Classical general relativity is insufficient in explaining the preservation or release of information contained in the infalling matter during Hawking radiation. Quantum mechanics suggest that Hawking radiation could be a type of thermal radiation, potentially leading to loss of information around the black hole. These circumstances give rise to fundamental inquiries on the characteristics the black hole thermodynamics including the destiny of information contained within them (Nottale, 1996).

2.7.3.4. Quest for a Quantum Signature

1. **The Cosmic Microwave Background (CMB):** Although it is difficult to directly observe quantum gravity effects, continuous cosmological measurements can offer indirect evidence and help shape theoretical advancements. Here are a few potential areas of inquiry. Deviation from a Gaussian pattern within the CMB data could potentially suggest the existence of fluctuations during inflation.

2. **Dark Matter and Dark Energy:** The enigmatic properties of dark matter with dark energy, which constitute most of the universe's energy content, continue to elude scientific understanding. Several ideas suggest that these mysterious entities may have beginnings rooted in quantum mechanics. Weakly Interacting huge particles (WIMPs), which are widely considered as potential dark matter candidates, are believed to be huge particles that emerge from the expansion of the standard theory like particle physics, incorporating aspects of supersymmetry.

The study of quantum mechanics in cosmology is an intriguing and demanding field that pushes the boundaries of our knowledge in our pursuit to understand the world. The pursuit theory of quantum gravity, which aims to reconcile the quantum or classical domains, is crucial for gaining a more profound comprehension of the origins, development, and ultimate destiny. Through the examination of theoretical frameworks, the search for observable indications using sophisticated cosmological tools, and the ongoing improvement of our comprehension of quantum physics, one persistently challenges the limits of understanding and strives toward a more comprehensive explanation of our world.

The pursuit of a theory of quantum gravity is an ongoing endeavor, characterized by various ideas vying for prominence. The objective of current research is to develop a comprehensive theory that unifies gravity with the other fundamental forces, while also providing a logical explanation for the early phases of the universe and the enigmatic phenomena observed in the quantum realm of gravity (Chaudhury et al., 2009).

2.7.4. Quantum Entanglement

Quantum entanglement, a fundamental concept in quantum physics, has fascinated physicists for many years due to its paradoxical nature and wide-ranging consequences.

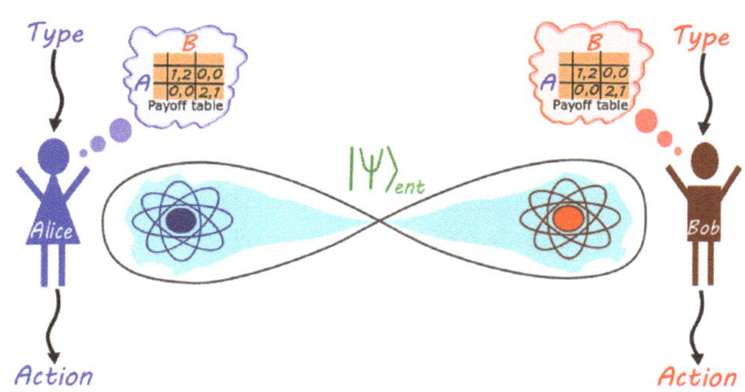

Figure 2.12. *Illustration of quantum entanglement,*

Source: Yihe Xue, Creative Commons License.

Entanglement is a phenomenon in which multiple quantum particles become intricately connected, sharing an identical quantum state despite their physical distance from each other. Essentially, the state of one interconnected particle is immediately linked with the other's, even over great distances. This seemingly contradictory relationship, widely referred to as "spooky action at a distance" from Albert Einstein, was confirmed through numerous experiments and offers significant potential for future technology (Laflorencie, 2016) (Figure 2.12).

2.7.4.1. The Essence of Entanglement

Now, examine two qubits, which are the basic building blocks for data in quantum computing. We will refer to them as qubit 1 and qubit 2. Within the framework of classical mechanics, a qubit can exist in one of two distinct states: 0 or 1. Yet, the allure of quantum mechanics resides in the concept of superposition. A qubit can concurrently exist in a superposition of its two states, which can be mathematically described as:

$$|\, Qubit\ State \rangle = \alpha|\, 0\rangle + \beta|\, 1\rangle$$

where α and β are complex numbers reflect the probability values of discovering the qubit in a particular state $|0\rangle or |1\rangle$, respectively, and $|\alpha|^2+|\beta|^2=1$.

Entanglement occurs when two qubits become uniquely interconnected uniquely. The entangled pair's overall state is represented by a unified wavefunction $(\Psi(q_1,q_2))$, which encompasses the states of both qubits. The Bell state is a straightforward example of an entangled state for two qubits:

$$|\Psi(q_1,q_2)\rangle=(1/\sqrt{2})(|00\rangle+|11\rangle)$$

Here, |00> represents both qubits being in the state |0>, and |11> represents both being in the state |1>. This entangled state indicates that the qubits correlate but the individual states remain uncertain until it is measured (Wootters, 1998).

2.7.4.2. Measurement and Correlation

The fascinating nature of entanglement becomes apparent when one of these entangled qubits is measured. Let's say one determines qubit 1 within the Bell state (Eq. 2). When a measurement is made, the wavefunction collapses and the qubit is forced into a specific state, either |0> or |1>. Almost magically, in the blink of an eye, qubit 2 undergoes a remarkable transformation, ensuring that the correlation remains intact. This connection remains intact regardless of the separation between the entangled qubits.

However, it's important to note that entanglement does not involve communication at speeds faster than light. When one qubit is measured, it doesn't send any information about its state to the other qubit. Instead, it merely uncovers the pre-existing correlation that is encoded into the entangled state (Margenau & Hill, 1961).

2.7.4.3. Applications of Quantum Entanglement

The remarkable characteristics of entanglement offer great promise for a wide range of technological advancements. Here are a few notable examples:

1. **Quantum Teleportation:** This fascinating accomplishment involves transferring the state of one qubit to another qubit located far away by utilizing entanglement and classical communication channels. Although it does not involve physically moving the particle, teleportation enables quantum information transfer with remarkable accuracy.

2. **Quantum Cryptography:** Entanglement offers a robust method for ensuring secure communication. By utilizing of entangled states, messages can be encoded to guarantee their integrity. Any attempt to intercept the communication would disturb the entanglement and be easily detectable.
3. **Quantum Computing:** Entanglement plays a crucial role in the foundation of quantum computers. Through the utilization of entangled qubits, quantum computers have the ability to solve problems that are beyond the capabilities of classical computers. This breakthrough has the potential to completely transform various fields, such as drug discovery, materials science, and financial modeling (Zeilinger, 1998).

2.7.4.4. Examples of Entanglement in Nature

Although commonly associated with experiments conducted in laboratories, entanglement extends beyond the realm of human creations. Naturally occurring phenomena can be observed in a variety of physical systems:

1. **Photons:** Nonlinear optical methods are commonly used to generate entangled photons. These interconnected pairs can be utilized in quantum communication protocols such as quantum key distribution.
2. **Atoms:** Trapped atoms in cold atom experiments have been effectively shown to be entangled. These systems provide advantageous platforms for investigating the fundamental characteristics of entanglement and its usesfor quantum information processing.
3. **Molecules:** Entanglement can also occur among spins of electrons in molecules. Gaining a comprehensive understanding of these complex interconnections is essential for the advancing the creation of innovative materials that possess specific characteristics (Somma et al., 2004).

2.7.4.5. Challenges and the Future of Entanglement Research

Utilizing entanglement for practical purposes poses numerous obstacles, despite its considerable potential:

1. **Scalability:** Establishing and preserving entanglement across a substantial number of qubits continues to be a major challenge. To fully harness the capabilities of quantum technology, it is imperative to find a solution to this limitation.
2. **Decoherence:** Entanglement is a delicate phenomenon that is easily affected by decoherence. Decoherence occurs when the entangled state interacts with the environment, resulting in the loss of coherence and transformation into a classical mixture of states. Practical applications heavily rely on mitigating decoherence, which can be achieved by through techniques such as error correction and fault-tolerant quantum computing.
3. **Long-Distance Distribution:** Preserving entanglement across extended distances presents an additional challenge. Factors such as defects in fiber optic cables for photons or faults in trapping

potential for atoms might decrease entanglement. Ongoing research is being conducted on the distribution of long-distance entanglement using quantum repeaters with satellite-based communication. Despite these challenges, the study of entanglement persists and thrives, propelled by its extraordinary potential. Here are some promising areas for future investigation.

4. **Quantum Networks:** Constructing expansive quantum networks that link physically distant quantum processors through entanglement is a crucial step toward distributed quantum computing and establishing a secure quantum communication infrastructure.
5. **Quantum Error Correction:** Developing resilient error correction techniques is crucial for preserving the integrity of entangled states and performing complex quantum computations. Researchers are actively investigating this field to overcome the limitations imposed by decoherence.
6. **Exploring New Forms of Entanglement:** By going beyond Bell states, the investigation and utilization of more complex forms of entanglement could result in innovative capabilities and applications in the field of quantum information processing. Exploration of topics such as graph states along with cluster states shows potential for future advancements.

The quest to fully exploit the potential of entanglement is ongoing. As researchers overcome obstacles and explore the intricacies of entanglement, it has the potential to transform several disciplines. This could lead to a new era of quantum technologies that have significant implications for science, communication, and computing (Erhard et al., 2020).

2.7.5. Applications of Quantum Mechanics in Technology

The ideas and mathematical framework of quantum mechanics have had a revolutionary impact in several technological fields. This section examines important applications, equations, and cases in which quantum mechanics is essential:

2.7.5.1. Quantum Computing

Quantum computing utilizes the fascinating principles of quantum mechanics to perform computations at an incredibly accelerated rate compared to traditional computers. The core equation that governs quantum computing is based on the concept of quantum bits and qubits, which are mathematically represented as:

$$|\psi = \alpha|0 + \beta|1$$

Here, $(|\alpha|^2 + |\beta|^2 = 1)$ Describes the likelihood amplitudes and manipulating these states by quantum gates results in intricate calculations (Figure 2.13).

Figure 2.13. *Schematic of a quantum computer.*

Source: Google Quantum AI, Creative Commons License.

- **Example:** Quantum algorithms, such as Shor's algorithm in factoring large numbers, have shown significant potential in surpassing classical algorithms. This has profound implications for fields such as cryptography and security (Rieffel and Polak, 2000).

2.7.5.2. Quantum Sensors

The field of quantum mechanics enables the creation of highly sensitive sensors, utilizing the fascinating phenomena of entanglement such as entanglement and interference. The sensitivity for these sensors is described by equations that involve the interaction between different states and external stimuli.

- **Example:** Quantum magnetometers harness the properties of atoms or ions to accurately measure magnetic fields, making them invaluable for applications such as medical imaging and geological exploration.

2.7.5.3. Quantum Communication

Quantum physics forms the foundation of secure communication protocols such as quantum key distribution (QKD), which ensures that information is highly resistant to eavesdropping. The equation governing QKD procedures involves the transfer of quantum states across a communication channel.

$$|\psi\rangle_{AB} = \frac{1}{\sqrt{2}}(|0\rangle_A|0\rangle_B + |1\rangle_A|1\rangle_B)$$

- **Example:** Quantum cryptography systems, like BB84, utilize the polarization state for photons to safely transfer cryptographic keys across extended distances (Ursin et al., 2007).

2.7.5.4. Quantum Metrology

Quantum mechanics enhances the accuracy of measurements beyond what is achievable with conventional methods, thereby influencing areas such as timekeeping, navigation, and the detection of gravitational waves. The equations that govern quantum metrology include Heisenberg's uncertainty principle and the quantum Fisher information.

- **Example:** Atomic clocks, which rely on the oscillations of atoms in superposed quantum states, can achieve accuracies of one second over billions of years. This level of precision is crucial for the functioning of global positioning systems (GPS) including satellite communication (Xiang & Guo, 2013).

2.7.5.5. Quantum Materials and Nanotechnology

Quantum mechanics is used to create and enhance innovative materials that with distinctive electrical, optical, and mechanical characteristics at the nanoscale. The equations that describe quantum confinement effects, including band structure estimates are essential in this discipline (Cava et al., 2021).

- **Example:** Quantum dots, small particles of semiconductors that exhibit quantum confinement phenomena, are employed in various applications such as displays, solar cells, and biological imaging because of their tunable electrical properties (Plotnitsky, 2011).

Ultimately, quantum mechanics not only enhances our understanding of fundamental physics, but also stimulates advancements in technology across many fields. These applications demonstrate how quantum mechanics can significantly advance modern technology by utilizing quantum phenomena and mathematical frameworks (Bauer et al., 2020).

ACTIVITY 2.1.

Simulate the behavior of a particle in a one-dimensional infinite potential well using computational tools or create a physical model to understand how the Schrödinger equation predicts energy quantization and wave function patterns within such a system.

SUMMARY

This chapter provides a comprehensive introduction to quantum physics, starting with the foundational principle of Heisenberg Uncertainty Principle. This principle is discussed in detail, including its applications to position and momentum, as well as energy and time. An illustrative example, the Heisenberg Microscope, is used to clarify these concepts.

Following this, the chapter explores wave-particle duality, highlighting the dual nature of particles and waves. This leads into a discussion on the wave function, its properties, and the fundamental postulates of quantum mechanics. The use of wave functions is explained, emphasizing their role in describing quantum systems.

The chapter then addresses the concepts of probability density and normalization of wave functions, providing a detailed understanding of how these functions describe the likelihood of finding particles in specific states.

A significant portion of the chapter is dedicated to the Schrödinger Wave Equation. The setup and derivation of the one-dimensional, time-independent Schrödinger equation are presented, along with examples include the infinite potential well and the particle in a box. The concepts of eigenvalues, eigenfunctions, orthogonality, and normalization are discussed in this context.

The applications of the Schrödinger Wave Equation are explored, including bound states in atoms and molecules, tunneling phenomena, quantum confinement, and scattering problems. These applications demonstrate the practical implications of the equation in various physical systems.

Advanced topics in quantum mechanics are introduced, such as the quantum harmonic oscillator, quantum tunneling in condensed matter physics, quantum mechanics in cosmology, and quantum entanglement. The chapter also discusses the technological applications of quantum mechanics, highlighting its relevance in modern science and technology.

REVIEW QUESTIONS

1. What is the Heisenberg uncertainty principle, and how does it define the limits of measurement in quantum mechanics?
2. Describe how the double-slit experiment demonstrates wave-particle duality.
3. How is probability density derived from the wave function?
4. Write down the time-independent Schrödinger wave equation and explain its terms.
5. Explain the concept of energy quantization in the context of bound states.
6. What is a quantum harmonic oscillator, and why is it significant in quantum mechanics?

7. Why is quantum tunneling important in the operation of devices like tunnel diodes and scanning tunneling microscopes?

MULTIPLE CHOICE QUESTIONS

1. **Heisenberg's Uncertainty Principle states that:**
   a. The position and momentum of a particle can be measured exactly at the same time.
   b. The position and momentum of a particle cannot both be precisely determined simultaneously.
   c. The energy and time of a particle can be measured exactly at the same time.
   d. The velocity and position of a particle cannot both be precisely determined simultaneously.

2. **Wave-particle duality refers to:**
   a. The phenomenon where particles exhibit only wave-like properties.
   b. The phenomenon where waves exhibit only particle-like properties.
   c. The phenomenon where particles exhibit both wave-like and particle-like properties.
   d. The phenomenon where particles and waves are indistinguishable.

7. **The Schrödinger Wave Equation is used to:**
   a. Describe the motion of macroscopic objects.
   b. Determine the energy levels of classical particles.
   c. Describe the quantum state of a system and predict the probability distribution of particles.
   d. Measure the uncertainty in position and momentum.

4. **In the context of the Schrödinger Wave Equation, eigenvalues represent:**
   a. The velocity of particles.
   b. The allowed energy levels of the quantum system.
   c. The uncertainty in measuring position and momentum.
   d. The probability density of finding a particle.

5. **Normalization of a wave function ensures that:**
   a. The wave function is always real-valued.
   b. The total probability of finding the particle within the given space is one.
   c. The wave function describes multiple particles simultaneously.
   d. The wave function has zero amplitude at infinity.

6. **Which of the following phenomena demonstrates quantum tunneling?**
   a. A particle moving in a straight line.

b. A particle passing through a barrier it classically should not be able to cross.
c. The reflection of light from a mirror.
d. The spreading of a wave packet over time.

7. Quantum entanglement refers to:

a. Particles that are physically connected and cannot be separated.
b. Particles that are independent and do not influence each other.
c. Particles that are in a state where the measurement of one particle instantaneously affects the state of another, regardless of the distance between them.
d. The uncertainty in the energy and time of a particle.

Answers to Multiple Questions

1. (b); 2. (c); 3. (c); 4. (b); 5. (b); 6. (b); 7. (c)

REFERENCES

1. Agravat, M. (2013). New methods for time correction of energy, momentum, and Heisenberg uncertainty principle. *Asian Journal of Science and Technology, 4*(5), 059.
2. Alexeev, Y., Amsler, M., Barroca, M. A., Bassini, S., Battelle, T., Camps, D., & Zubarev, D. (2024). Quantum-centric supercomputing for materials science: A perspective on challenges and future directions. *Future Generation Computer Systems, 160*, 666–710.
3. Alhaidari, A. D. (2019). Representation of the quantum mechanical wavefunction by orthogonal polynomials in the energy and physical parameters. *Communications in Theoretical Physics, 72*(1), 015104.
4. Alvarez, G., Graffi, S., & Silverstone, H. J. (1988). Transition from classical mechanics to quantum mechanics: x^4 perturbed harmonic oscillator. *Physical Review A, 38*(4), 1687.
5. Amico, L., Anderson, D., Boshier, M., Brantut, J. P., Kwek, L. C., Minguzzi, A., & von Klitzing, W. (2022). Colloquium: Atomtronic circuits: From many-body physics to quantum technologies. *Reviews of Modern Physics, 94*(4), 041001.
6. Arndt, M., Juffmann, T., & Vedral, V. (2009). Quantum physics meets biology. *HFSP Journal, 3*(6), 386–400.
7. Baily, C., & Finkelstein, N. D. (2015). Teaching quantum interpretations: Revisiting the goals and practices of introductory quantum physics courses. *Physical Review Special Topics—Physics Education Research, 11*(2), 020124.
8. Bâldea, I., Köppel, H., & Cederbaum, L. S. (2001). Collective quantum tunneling of strongly correlated electrons in commensurate mesoscopic rings. *The European Physical Journal B—Condensed Matter and Complex Systems, 20*, 289–299.
9. Barbosa, T. N., Almeida, M. M., & Prudente, F. V. (2015). A quantum Monte Carlo study of confined quantum systems: Application to harmonic oscillator and

hydrogenic-like atoms. *Journal of Physics B: Atomic, Molecular and Optical Physics, 48*(5), 055002.

10. Barukčić, I. (2016). Anti Heisenberg—The end of Heisenberg's uncertainty principle. *Journal of Applied Mathematics and Physics, 4*(5), 881–887.
11. Bastardis, R., Guihéry, N., & de Graaf, C. (2007). Microscopic origin of isotropic non-Heisenberg behavior in S=1 magnetic systems. *Physical Review B—Condensed Matter and Materials Physics, 76*(13), 132412.
12. Bauer, B., Bravyi, S., Motta, M., & Chan, G. K. L. (2020). Quantum algorithms for quantum chemistry and quantum materials science. *Chemical Reviews, 120*(22), 12685–12717.
13. Berkdemir, A., Berkdemir, C., & Sever, R. (2006). Eigenvalues and eigenfunctions of Woods–Saxon potential in PT-symmetric quantum mechanics. *Modern Physics Letters A, 21*(27), 2087–2097.
14. Biswas, S., Modak, B., & Biswas, D. (1997). Wave function of the universe from complex minisuperspace trajectories. *Physical Review D, 55*(8), 4673.
15. Bittner, E. R. (2000). Quantum tunneling dynamics using hydrodynamic trajectories. *The Journal of Chemical Physics, 112*(22), 9703–9710.
16. Bunge, M. (2003). Velocity operators and time-energy relations in relativistic quantum mechanics. *International Journal of Theoretical Physics, 42*, 135–142.
17. Busch, P., Grabowski, M., & Lahti, P. J. (2009). *Operational Quantum Physics*. Springer Science & Business Media.
18. Busch, P., Heinonen, T., & Lahti, P. (2007). Heisenberg's uncertainty principle. *Physics Reports, 452*(6), 155–176.
19. Carosso, A. (2022). Quantization: History and problems. *Studies in History and Philosophy of Science, 96*, 35–50.
20. Castro, A. S. D. (2020). Frustrating use of the Laplace transform for the quantum states of a particle in a box. *Revista Brasileira de Ensino de Física, 42*, 20200079.
21. Cava, R., De Leon, N., & Xie, W. (2021). Introduction: Quantum materials. *Chemical Reviews, 121*(5), 2777–2779.
22. Cetto, A. M., la Peña, L. D., & Valdés-Hernández, A. (2015). Specificity of the Schrödinger equation. *Quantum Studies: Mathematics and Foundations, 2*(3), 275–287.
23. Chang, D. C. (2021). Review on the physical basis of wave–particle duality: Conceptual connection between quantum mechanics and the Maxwell theory. *Modern Physics Letters B*, *35*(13), 2130004.
24. Chaudhury, S., Smith, A., Anderson, B. E., Ghose, S., & Jessen, P. S. (2009). Quantum signatures of chaos in a kicked top. *Nature*, *461*(7265), 768–771.
25. Chen, E. K. (2019). Realism about the wave function. *Philosophy Compass*, *14*(7), e12611.
26. Christiansen, O. (2007). Vibrational structure theory: New vibrational wave function methods for calculation of anharmonic vibrational energies and vibrational contributions to molecular properties. *Physical Chemistry Chemical Physics*, *9*(23), 2942–2953.

27. Cottet, A., Dartiailh, M. C., Desjardins, M. M., Cubaynes, T., Contamin, L. C., Delbecq, M., & Kontos, T. (2017). Cavity QED with hybrid nanocircuits: From atomic-like physics to condensed matter phenomena. *Journal of Physics: Condensed Matter*, *29*(43), 433002.

28. Dekker, H. (1981). Classical and quantum mechanics of the damped harmonic oscillator. *Physics Reports*, *80*(1), 1–110.

29. Duo, S., & Zhang, Y. (2015). Computing the ground and first excited states of the fractional Schrödinger equation in an infinite potential well. *Communications in Computational Physics*, *18*(2), 321–350.

30. Dutta Majumder, D. K., & Dutta, S. K. (2007). A new look at Heisenberg's uncertainty principle: A cybernetics and general dynamical systems approach. *Kybernetes*, *36*(5/6), 754–767.

31. Dysthe, K. B. (1979). Note on a modification to the nonlinear Schrödinger equation for application to deep water waves. *Proceedings of the Royal Society of London. A. Mathematical and Physical Sciences*, *369*(1736), 105–114.

32. Eckhardt, B., Fishman, S., Keating, J., Agam, O., Main, J., & Müller, K. (1995). Approach to ergodicity in quantum wave functions. *Physical Review E*, *52*(6), 5893.

33. Erhard, M., Krenn, M., & Zeilinger, A. (2020). Advances in high-dimensional quantum entanglement. *Nature Reviews Physics*, *2*(7), 365–381.

34. Erol, M. (2010). Schrödinger wave equation and function: Basics and concise relations with consciousness/mind. *NeuroQuantology*, *8*(1), 98–100.

35. Faddeev, L. D. (1990). Lectures on quantum inverse scattering method. In *New Problems, Methods and Techniques in Quantum Field Theory and Statistical Mechanics. Series on Advances in Statistical Mechanics* (Vol. 6, pp. 7–54).

36. Flatté, S. M., Bracher, C., & Wang, G. Y. (1994). Probability-density functions of irradiance for waves in atmospheric turbulence calculated by numerical simulation. *JOSA A*, *11*(7), 2080–2092.

37. Franke-Arnold, S., Barnett, S. M., Yao, E., Leach, J., Courtial, J., & Padgett, M. (2004). Uncertainty principle for angular position and angular momentum. *New Journal of Physics*, *6*(1), 103.

38. Fröman, N. (1978). Phase-integral formulas for level densities, normalization factors, and quantal expectation values, not involving wave functions. *Physical Review A*, *17*(2), 493.

39. Gaeta, Z. D., & Stroud Jr, C. R. (1990). Classical and quantum-mechanical dynamics of a quasiclassical state of the hydrogen atom. *Physical Review A*, *42*(11), 6308.

40. Gell-Mann, M., & Hartle, J. B. (1996). Quantum mechanics in the light of quantum cosmology. In *Foundations of Quantum Mechanics in the Light of New Technology: Selected Papers from the Proceedings of the First through Fourth International Symposia on Foundations of Quantum Mechanics* (Vol. 1, pp. 347–369).

41. Gill, S. S., Kumar, A., Singh, H., Singh, M., Kaur, K., Usman, M., & Buyya, R. (2022). Quantum computing: A taxonomy, systematic review and future directions. *Software: Practice and Experience*, *52*(1), 66–114.

42. Gopakumar, R. (2015). String theory and the conundrums of quantum gravity. *Current Science*, *8*, 2265–2270.

43. Gros, C. (1989). Physics of projected wavefunctions. *Annals of Physics*, *189*(1), 53–88.

44. Habibi, H., Zeuthen, E., Ghanaatshoar, M., & Hammerer, K. (2016). Quantum feedback cooling of a mechanical oscillator using variational measurements: Tweaking Heisenberg's microscope. *Journal of Optics*, *18*(8), 084004.

45. Hadzidaki, P. (2008). The Heisenberg Microscope: A powerful instructional tool for promoting meta-cognitive and meta-scientific thinking on quantum mechanics and the nature of science. *Science & Education*, *17*, 613–639.

46. Helal, M. A., & Seadawy, A. R. (2009). Variational method for the derivative nonlinear Schrödinger equation with computational applications. *Physica Scripta*, *80*(3), 035004.

47. Hilgevoord, J. (1996). The uncertainty principle for energy and time. I. *American Journal of Physics*, *64*(12), 1451–1456.

48. Hilgevoord, J. (1998). The uncertainty principle for energy and time. II. *American Journal of Physics*, *66*(5), 396–402.

49. Hioe, F. T., & Montroll, E. W. (1975). Quantum theory of anharmonic oscillators. I. Energy levels of oscillators with positive quartic anharmonicity. *Journal of Mathematical Physics*, *16*(9), 1945–1955.

50. Hofstadter, D. R. (1976). Energy levels and wave functions of Bloch electrons in rational and irrational magnetic fields. *Physical Review B*, *14*(6), 2239.

51. Höhn, P. A. (2019). Switching internal times and a new perspective on the 'wave function of the universe'. *Universe*, *5*(5), 116.

52. Isakov, S. V., Mazzola, G., Smelyanskiy, V. N., Jiang, Z., Boixo, S., Neven, H., & Troyer, M. (2016). Understanding quantum tunneling through quantum Monte Carlo simulations. *Physical Review Letters*, *117*(18), 180402.

53. Jensen, R. V., Susskind, S. M., & Sanders, M. M. (1991). Chaotic ionization of highly excited hydrogen atoms: Comparison of classical and quantum theory with experiment. *Physics Reports*, *201*(1), 1–56.

54. Kastrup, H. A. (2007). A new look at the quantum mechanics of the harmonic oscillator. *Annalen der Physik*, *519*(7, 8), 439–528.

55. Kavka, J. J., Kerbrat, D., & Shegelski, M. R. (2010). Quantum tunneling and reflection of a molecule with a single bound state. *Physical Review A: Atomic, Molecular, and Optical Physics*, *81*(2), 022708.

56. Kijowski, J. (1974). On the time operator in quantum mechanics and the Heisenberg uncertainty relation for energy and time. *Reports on Mathematical Physics*, *6*(3), 361–386.

57. Kim, S. K., Yang, J. Y., & Oh, J. (2006). Q-values for P and S waves in the southern Korean Peninsula based on the coda-normalization method. *Geosciences Journal*, *10*, 465–477.

58. Krenn, M., & Zeilinger, A. (2020). Predicting research trends with semantic and

neural networks with an application in quantum physics. *Proceedings of the National Academy of Sciences*, *117*(4), 1910–1916.

59. Laflorencie, N. (2016). Quantum entanglement in condensed matter systems. *Physics Reports*, *646*, 1–59.
60. Leonhardt, U. (2003). Quantum physics of simple optical instruments. *Reports on Progress in Physics*, *66*(7), 1207.
61. Lewis, G. N., & Mayer, J. E. (1929). The quantum laws and the uncertainty principle of Heisenberg. *Proceedings of the National Academy of Sciences*, *15*(2), 127–139.
62. Liu, Y. X., Özdemir, Ş. K., Miranowicz, A., & Imoto, N. (2004). Kraus representation of a damped harmonic oscillator and its application. *Physical Review A: Atomic, Molecular, and Optical Physics*, *70*(4), 042308.
63. Lundeen, J. S., Sutherland, B., Patel, A., Stewart, C., & Bamber, C. (2011). Direct measurement of the quantum wavefunction. *Nature*, *474*(7350), 188–191.
64. Maor, E. (2009). How I got to understand Heisenberg's uncertainty principle. *Math Horizons*, *16*(4), 5–33.
65. Margenau, H., & Hill, R. N. (1961). Correlation between measurements in quantum theory. *Progress of Theoretical Physics*, *26*(5), 722–738.
66. Marshman, E., & Singh, C. (2019). Validation and administration of a conceptual survey on the formalism and postulates of quantum mechanics. *Physical Review Physics Education Research*, *15*(2), 020128.
67. Mohebbi, A., Abbaszadeh, M., & Dehghan, M. (2013). The use of a meshless technique based on collocation and radial basis functions for solving the time fractional nonlinear Schrödinger equation arising in quantum mechanics. *Engineering Analysis with Boundary Elements*, *37*(2), 475–485.
68. Nelson, E. (1966). Derivation of the Schrödinger equation from Newtonian mechanics. *Physical Review*, *150*(4), 1079.
69. Nomura, Y. (2021). From the black hole conundrum to the structure of quantum gravity. *Modern Physics Letters A*, *36*(8), 2130007.
70. Nottale, L. (1996). Scale relativity and fractal space-time: Applications to quantum physics, cosmology and chaotic systems. *Chaos, Solitons & Fractals*, *7*(6), 877–938.
71. Ord, G. N. (2009). Quantum mechanics in a two-dimensional spacetime: What is a wavefunction? *Annals of Physics*, *324*(6), 1211–1218.
72. Owerre, S. A., & Paranjape, M. B. (2015). Macroscopic quantum tunneling and quantum–classical phase transitions of the escape rate in large spin systems. *Physics Reports*, *546*, 1–60.
73. Ozawa, M. (2015). Heisenberg's original derivation of the uncertainty principle and its universally valid reformulations. *Current Science*, *8*(3), 2006–2016.
74. Park, S., Brooks, F. J., Villa, U., Su, R., Anastasio, M. A., & Oraevsky, A. A. (2022). Normalization of optical fluence distribution for three-dimensional functional optoacoustic tomography of the breast. *Journal of Biomedical Optics*, *27*(3), 036001.
75. Pattanayak, A. K., & Schieve, W. C. (1994). Gaussian wave-packet dynamics:

Semiquantal and semiclassical phase-space formalism. *Physical Review E*, *50*(5), 3601.

76. Piron, C. (1976). On the foundations of quantum physics. In *Quantum Mechanics, Determinism, Causality, and Particles: An International Collection of Contributions in Honor of Louis de Broglie on the Occasion of the Jubilee of His Celebrated Thesis* (Vol. 2, pp. 105–116). Dordrecht: Springer Netherlands.
77. Plotnitsky, A. (2011). "Dark materials to create more worlds": On causality in classical physics, quantum physics, and nanophysics. *Journal of Computational and Theoretical Nanoscience*, *8*(6), 983–997.
78. Prasad, N. S., & Roychoudhuri, C. (2011). Microscope and spectroscope results are not limited by Heisenberg's uncertainty principle. In *The Nature of Light: What are Photons? IV*. SPIE (Vol. 8121, pp. 162–168).
79. Rashkovskiy, S. A. (2023). Quantization of Hamiltonian and non-Hamiltonian systems. *Communications in Analysis and Mechanics*, *15*(2), 267–288.
80. Raymer, M. G. (1997). Measuring the quantum mechanical wave function. *Contemporary Physics*, *38*(5), 343–355.
81. Rempe, G. (2000). Quantum physics of entangled systems: Wave–particle duality and atom–photon molecules. *Annalen der Physik*, *512*(11, 12), 843–850.
82. Rieffel, E., & Polak, W. (2000). An introduction to quantum computing for non-physicists. *ACM Computing Surveys (CSUR)*, *32*(3), 300–335.
83. Ringbauer, M., Duffus, B., Branciard, C., Cavalcanti, E. G., White, A. G., & Fedrizzi, A. (2015). Measurements on the reality of the wavefunction. *Nature Physics*, *11*(3), 249–254.
84. Rosner, W., Wunner, G., Herold, H., & Ruder, H. (1984). Hydrogen atoms in arbitrary magnetic fields. I. Energy levels and wavefunctions. *Journal of Physics B: Atomic and Molecular Physics*, *17*(1), 29.
85. Sebens, C. T. (2019). Electromagnetism as quantum physics. *Foundations of Physics*, *49*, 365–389.
86. Shirokov, M. I. (1978). Signal velocity in quantum electrodynamics. *Soviet Physics Uspekhi*, *21*(4), 345.
87. Silverman, B. W. (1982). On the estimation of a probability density function by the maximum penalized likelihood method. *The Annals of Statistics*, *7*(1), 795–810.
88. Somma, R., Ortiz, G., Barnum, H., Knill, E., & Viola, L. (2004). Nature and measure of entanglement in quantum phase transitions. *Physical Review A: Atomic, Molecular, and Optical Physics*, *70*(4), 042311.
89. Stenholm, S. (1973). Quantum theory of electromagnetic fields interacting with atoms and molecules. *Physics Reports*, *6*(1), 1–121.
90. Stucky, G. D., & Mac Dougall, J. E. (1990). Quantum confinement and host/guest chemistry: Probing a new dimension. *Science*, *247*(4943), 669–678.
91. Tannous, C., Fakhreddine, K., & Langlois, J. (2008). The canonical function method and its applications in quantum physics. *Physics Reports*, *467*(6), 173–204.

92. Ursin, R., Tiefenbacher, F., Schmitt-Manderbach, T., Weier, H., Scheidl, T., Lindenthal, M., & Zeilinger, A. (2007). Entanglement-based quantum communication over 144 km. *Nature Physics*, *3*(7), 481–486.

93. Vergel, D. G., & Villaseñor, E. J. (2009). The time-dependent quantum harmonic oscillator revisited: Applications to quantum field theory. *Annals of Physics*, *324*(6), 1360–1385.

94. Wagner, F. (2021). Generalized uncertainty principle or curved momentum space? *Physical Review D*, *104*(12), 126010.

95. Wang, H. Y. (2022). There is no vacuum zero-point energy in our universe for massive particles within the scope of relativistic quantum mechanics. *Physics Essays*, *35*(3), 270–275.

96. Weinacht, T. C., Ahn, J., & Bucksbaum, P. H. (1999). Controlling the shape of a quantum wavefunction. *Nature*, *397*(6716), 233–235.

97. Werner, R. F., & Farrelly, T. (2019). Uncertainty from Heisenberg to today. *Foundations of Physics*, *49*, 460–491.

98. Wootters, W. K. (1998). Quantum entanglement as a quantifiable resource. *Philosophical Transactions of the Royal Society of London. Series A: Mathematical, Physical and Engineering Sciences*, *356*(1743), 1717–1731.

99. Xiang, G. Y., & Guo, G. C. (2013). Quantum metrology. *Chinese Physics B*, *22*(11), 110601.

100. Yang, B., & Zhou, A. (2021). Eigenfunction behavior and adaptive finite element approximations of nonlinear eigenvalue problems in quantum physics. *ESAIM: Mathematical Modelling and Numerical Analysis*, *55*(1), 209–227.

101. Zeilinger, A. (1998). Quantum entanglement: A fundamental concept finding its applications. *Physica Scripta*, *1998*(76), 203.

102. Zeilinger, A. (1999). Experiment and the foundations of quantum physics. *Reviews of Modern Physics*, *71*(2), S288.

103. Zheng, C. (2007). A perfectly matched layer approach to the nonlinear Schrödinger wave equations. *Journal of Computational Physics*, *227*(1), 537–556.

104. Zhou, E. L., Wu, Y., Lin, X. Y., Li, Q. Q., & Xiang, Y. (2021). A normalization strategy for BESO-based structural optimization and its application to frequency response suppression. *Acta Mechanica*, *232*, 1307–1327.

CHAPTER 3

Conducting Materials

LEARNING OBJECTIVES

After studying this chapter, you will be able to:

1. Understand the classical free electron theory and its application to electrical and thermal conductivity.
2. Explain Wiedemann-Franz law and its relevance in conducting materials.
3. Analyze the Fermi-Dirac distribution function and the concept of density of energy states.
4. Describe the properties, types, and carrier concentrations of semiconducting materials.
5. Explore the Hall effect and its applications, including the expression of Hall coefficient and Hall voltage.
6. Discuss the classification, domain theory, and magnetization process of magnetic materials.
7. Understand the principles, types, and applications of superconductors, including BCS theory and high Tc superconductors.
8. Examine the properties and applications of dielectric materials, metallic glasses, shape memory alloys, and nanomaterials.
9. Understand the classification and applications of biomaterials in various fields such as ophthalmology and dentistry.

KEY TERMS FROM THIS CHAPTER

Classical free electron theory
Electrical conductivity
Ferromagnetic domain theory
Shape memory alloys
Density of energy states
Fermi-Dirac distribution function
Hall effect
Superconductors

UNIT INTRODUCTION

Electrical conductivity refers to the capacity of a substance to efficiently transmit an electric current. This phenomenon results from a movement of particles with electrical charges within the structure of the material. Significantly, the existence of "free electrons," which are not bound to atoms, has a decisive influence. These unbound electrons act as mobile charge carriers, enabling the transfer of current when a potential difference (the voltage) is supplied to the substance (Wan, 2008).

Materials that have a high density of unbound electrons demonstrate exceptional conductivity and serve as the basis for a wide range of scientific and technological uses. These applications encompass the effective transfer of electrical signals, the provision of power to devices, and the construction of complex circuits (Pal et al., 2020).

When it comes to electrical conductivity, metals are the clear winners among different material types. Their atomic configuration enables the presence of a dispersed group of unbound electrons, which can easily move in response to an electric field. Metals are preferred for electrical cables, parts, and heat sinks due to their superior electrical conduction. Typical examples include aluminum, copper, gold, and silver (Dubey and Leclerc, 2011).

Notably, there is a strong correlation between thermal and electrical conductivity. Several materials that possess excellent electrical conductivity also exhibit effective heat transfer. This is because heat can be conducted by the motion of particles with charge within a substance. As a result, a metallic wire that effectively carries electrical current will also easily transfer heat energy effectively (Tallman et al., 2002).

3.1. CLASSICAL FREE ELECTRON THEORY

The free electron gas theory is based on the following assumptions:

1. A metal contains many unbound electrons that may travel freely throughout its volume, similar to gas molecules in a container. Randomly moving free electrons collide with stationary positive ions in the lattice structure or with other free electrons. All electron collisions are elastic; thus, no energy is lost.
2. Free electrons follow the conventional Maxwell-Boltzmann statistics on their velocity and energy distribution (Li and Yu, 2011).
3. The free electrons are moving within a uniformly distributed potential field generated by the immobile ions in the lattice.
4. In the absence of an electric field, the free electrons travel randomly in all directions with equal probability, causing the current density vector to be zero (Mills, 2001).
5. When a metallic substance is exposed to an external electrical field, its electrons drift in the direction opposite to the field. The drift velocity is the average speed of this movement. Drift velocity, along with stochastic velocity. The drift velocity controls the electric current in metals (Sarachik and Schappert, 1970).

Below are a few fundamental definitions of terminology commonly used in the field of materials science:

1. **Drift Velocity:** The average velocity of the unbound electrons as they move towards the positive terminal in response to the electric field.
2. **Mobility:** The term refers to the velocity at which the charge carrier moves in response to an applied electric field, measured per unit of the field's strength (Luong and Shaw, 1971).
3. **Collision Time:** The average time that a free electron spends between two consecutive collisions.
4. **Mean Free Path:** The mean free route refers to the average distance covered when a free electron moves before it has a collision.
5. **Relaxation Time:** The electron's relaxation time refers to the duration it takes for a free electron to return to its original position after being displaced in the vicinity of an electric field (Duboz et al., 1995).

3.1.1. Electrical Conductivity

Electrical conductivity, represented by the symbol σ, is the quantitative measure of how easily electric charge can move through a conductor. The current in a conductor is measured by the rate of charge flow through a specific area, in relation to the applied potential difference (voltage).

$$\sigma = \frac{J}{E}$$

The unit of measurement is ohm per meter ($\Omega\, m^{-1}$) or Siemens per meter (Sm^{-1}).

Remember

Electrical conductivity in materials depends on the mobility of free electrons and the density of states available for conduction.

The mathematical expression for electric conductivity is based on the movement of an electron with charge 'e' in a metal rod when subjected to an electric field (E). The electron moves in the direction opposite to the applied field with a velocity v_d. The velocity that is utilized to characterize this phenomenon is referred to as drift velocity (Radzuan et al., 2017).

The Lorentz force exerted on the electron:

$$F = eE \quad (3.1)$$

The force in question is commonly referred to as the electron's driving force (Figure 3.1).

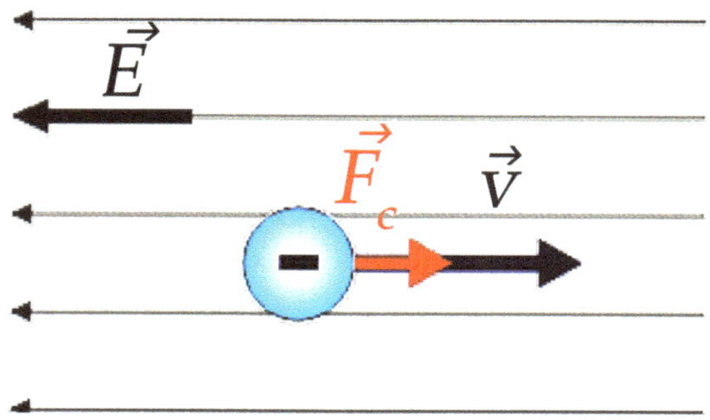

Figure 3.1. *Electrons moving in a uniform field.*

Source: Physics Tasks; Creative Commons License.

As a result of this force, each electron experiences an acceleration of 'a.'

According to Newton's second principle of motion:

$$F = ma \quad (3.2)$$

From Equations 3.1 and 3.2:

$$ma = eE$$

or

$$a = \frac{-eE}{m} \quad (3.3)$$

The interest is in formulating the equation that relates drift velocity (υ_d), acceleration (a), and relaxation time (τ). Presented below is the text in a simple and plain format (Kerner, 1956):

$$Acceleration(a) = \frac{Drift\ velocity}{Relaxation\ time}$$

$$a = \frac{\upsilon_d}{\tau}$$

$$\upsilon_d = a\ \tau \quad (3.4)$$

Substituting Equations 3.3 in 3.4:

$$\nu_d = \frac{e\tau}{m}E \quad (3.5)$$

According to Ohms' law, the current density (J) is stated as:

$$J = \sigma E \text{ or } \sigma = \frac{J}{E} \quad (3.6)$$

Here, σ represents the electron's electrical conductivity. However, the expression for current density in relation to drift velocity is provided as:

$$J = nev_d \quad (3.7)$$

Substituting Equations 3.5 in 3.7:

$$\frac{e\tau}{m} = ne\frac{E}{J} = \frac{ne^2\tau}{m} \quad (3.8)$$

On comparing Equations 3.6 and 3.8 (Friedman, 2005),

$$\sigma = \frac{ne^2\tau}{m} \quad (3.9)$$

3.1.2. Thermal Conductivity

Thermal conductivity, represented by the symbol K, measures the rate at which heat is transferred through a material with a specific area and temperature difference:

$$Q = K\frac{dT}{dx}$$

The thermal conductivity formula for a metal is: Two evenly spaced cross-sections, A and B will be analyzed, within a metallic rod AB, separated by a distance λ. Consider A at a higher temperature and B at a little lower temperature (T-dT). Heat transfer occurs from both A and B through the motion of electrons. The quantity n represents the density of conduction electrons, whereas v denotes their average velocity. Collisions arise when electrons undergo movement within the rod. As a result, the electrons near point A have a reduction in their kinetic energy, while the electrons near point B have a rise in their kinetic energy (Burger et al., 2016) (Figure 3.2).

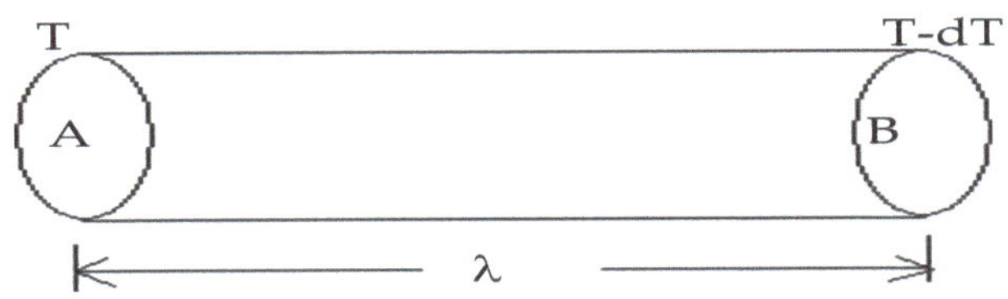

Figure 3.2. *Metallic rod AB separated by a distance λ.*

Source: JECRC Foundation; Creative Commons License.

The mean kinetic energy associated with an electron at point A $= \frac{3}{2}KT$ at B, it is $\frac{3}{2}K(T-dT)$. The surplus of kinetic energy transported from point A to point B equals:

$$\frac{3}{2}kdT \quad (3.10)$$

The electron flux from points A to B is:

$$\frac{3}{2}n\upsilon$$

The energy flux is the measure of energy moved from point A to point B per unit area and unit time, is:

$$1/4n\upsilon kdT \qquad (3.11)$$

Similarly, the energy flux from B to A by area by unit time is (Carruthers, 1961):

$$= -\frac{1}{4}n\upsilon kdT \qquad (3.12)$$

Therefore, the total energy carried from A to B for each unit area per period is:

$$Q = -\frac{1}{2}n\upsilon kdT \qquad (3.13)$$

However, according to the fundamental concept of thermal conductivity, the quantity of heat transmitted per unit area per time is:

$$Q = K\frac{dT}{\lambda}, \text{ that is, } \frac{1}{2}n\upsilon kdT = K\frac{dT}{\lambda}; K = \frac{1}{2}n\upsilon k\lambda \qquad (3.14)$$

For metals:

$$\tau = \tau_c = \frac{\lambda}{\upsilon} = \tau\upsilon = \lambda \qquad (3.15)$$

Putting Equations 3.14 in 3.15 (Cahill, 1990),

$$K = \frac{1}{2}n\upsilon^2 k\tau \qquad (3.16)$$

3.1.3. Wiedemann–Franz Law

Both the thermal and electric conductivity of a metal are directly correlated with the metal's absolute temperature, as stated by the law (Ahmad and Mahanti, 2010).

$$\sigma = \frac{ne^2\tau}{m}$$

and

$$K = \frac{1}{2}n\upsilon^2 k\tau$$

$$\frac{\text{Thermal conductivity}}{\text{Electrical conductivity}} \aleph \frac{k}{\sigma} \quad \frac{\frac{1}{2}\pm^2 \tau}{\frac{ne^2\tau}{m}}; \frac{k}{\sigma} \quad \frac{1}{2}\frac{mv^2k}{e^2} \qquad (3.17)$$

The K.E. associated with an electron is:

$$\frac{1}{2}mv^2 = \frac{3}{2}kT \qquad (3.18)$$

Substituting Equations 3.18 in 3.17:

$$\frac{k}{\sigma} = \frac{3}{2}\frac{kT \times k}{e^2} = \frac{3}{2}\frac{k^2T}{e^2}, or \frac{k}{\sigma} = \frac{3}{2}\left[\frac{k}{2}\right]^2 T, or \frac{k}{\sigma} = LT \qquad (3.19)$$

Whereas $L = \frac{3}{2}\left[\frac{k}{e}\right]^2$ is called the Lorentz constant.

$$\frac{k}{\sigma}\alpha T \qquad (3.20)$$

Therefore, it has been demonstrated that the thermal conductivity ratio of a metal has a direct connection with the metal's absolute temperature (Völklein et al., 2009).

3.1.4. Fermi-Dirac Distribution Function

3.1.4.1. Fermi Function F (E)

The Fermi-Dirac distribution function quantifies the likelihood of an electron's presence in a certain energy state at a given temperature.

$$F(E) = \frac{1}{1+e^{(E-E_F)/K_BT}}$$

where; K_B is Boltzmann's constant (Justice et al., 2019).

3.1.4.2. Effect of Temperature on Fermi Function

- **Case (i):** The probability $E < E_F$ at absolute zero temperature.

$$F(E) = \frac{1}{1+e^{(E-E_F)/KT}}$$

or

$$F(E) = \frac{1}{1+e^{-\infty} = 0} = \frac{1}{1+0} = \frac{1}{1}\left[\because e^{-\infty} = 0\right]; \therefore F(E) = 1$$

Therefore, at a temperature of 0 Kelvin, there is a 100% probability for the electrons to be at energy levels that are lower than the Fermi level (Park and Donnet, 1998) (Figure 3.3).

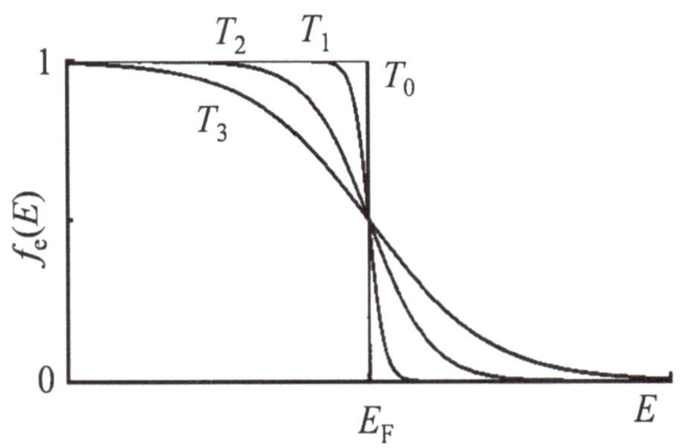

Figure 3.3. *Fermi-Dirac distribution function at different temperatures: $T_3 > T_2 > T_1$ (and $T_0 = 0$ K). At the absolute zero temperature (T_0), the probability of an electron to have an energy below the Fermi energy EF is equal to 1, while the probability to have higher energy is zero.*

Source: Pradip Basnet; Creative Commons License.

- **Case (ii):** Probability for $E > E_F$ at zero Kelvin.

  It is,

$$F(E) = \frac{1}{1+e^{(+ve/0)}} = \frac{1}{1+e^{\infty}} = \frac{1}{1+\infty} = \frac{1}{\infty} = 0 \therefore F(E) = 0$$

Hence, the probability of electrons occupying energy levels beyond the Fermi energy level is absolutely zero. Based on the two situations mentioned above, when the temperature is at absolute zero (T = 0 K), the function $F(E)$ exhibits a step-like fluctuation for different energy levels (Figure 3.4).

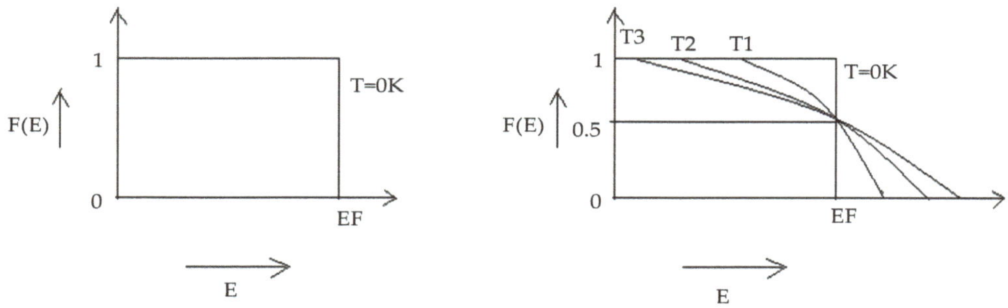

Figure 3.4. *Variation of Fermi distribution function with E at different temperatures.*

Source: NIT Srinagar; Creative Commons License.

- **Case (iii): Probability at Normal Temperature:** At room temperature, the probability decreases from a value of 1 as the energy values approach but remain slightly below than the Fermi energy (E_F). As the temperature increases beyond absolute zero (0 K), the Fermi function $F(E)$ changes in accordance with the energy E.

  For any temperature that is not equal to absolute zero and when the energy is equal to the Fermi energy (E_F),

  $$F(E) = \frac{1}{1+e^0} = \frac{1}{1+1} = \frac{1}{2} = 50\%$$

  Therefore, there is a 50% probability of the electrons to be at the Fermi level. Moreover, when the value of E is greater than E_F, the likelihood decreases rapidly and approaches zero.

- **Case (iv): Probability at High Temperature:** When the value of the product of the temperature (kT) and the Fermi energy (E_F) greatly exceeds 1, the behavior of electrons no longer follows the principles of quantum mechanics, and the Fermi distribution function can be approximated by the ordinary Boltzmann distribution (McDougall and Stoner, 1938).

3.1.5. Density of Energy States

The term refers to the quantity of electron states that are accessible within a certain volume for a specific range of energy, from E to $E+dE$. The symbol $Z(E)dE$ represents it (Figure 3.5).

Consider a cubic object with sides of length 'a.' In three dimensions, the quantum numbers *nx*, *ny*, and *nz* define 'a' the spatial coordinates within the cube.

A radius vector (n) extends from the origin 'O' to the dimensions (*nx*, *ny*, *nz*). Every point on the sphere has the same energy level, E. The diameter (or radius) of the sphere with the energy E is given by the equation $n^2 = nx^2 + ny^2 + nz^2$. The sphere contains numerous layers. Each layer is associated with a unique combination of *nx*, *ny*, and *nz*. Thus, it reflects a radius-specific energy value. In this case, "unit volume" means a single energy state (Blakemore, 1982).

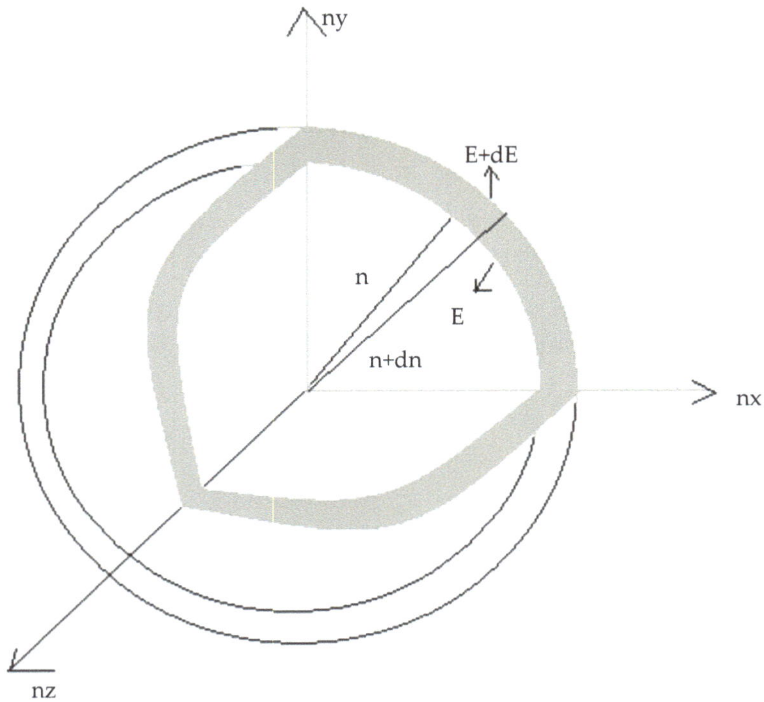

Figure 3.5. *Illustration of energy of states.*

Source: NIT Srinagar; Creative Commons License.

The formula for the total amount of levels of energy in a sphere with a radius of 'n' is 4/3 πn^3.

Since *nx*, *ny*, and *nz* may only be positive integers, concentrate on 1/8th of the spherical volume, or 1/8th of the sphere. With a radius of 'n' and an energy level of E, one-eighth of a sphere may reach the following energy levels:

$$=\frac{1}{8}\left[\frac{4}{3}\pi n^3\right]$$

Likewise, the number of energy states that can be found within a specific portion of a sphere with a radius of '±+ ' that corresponds to an energy level of $E+dE$:

$$=\frac{1}{8}\left[\frac{4}{3}\pi(n+dn)^3\right]$$

The number of energy levels that can be observed under the shell with a radius of n and $n + dn$ are currently analyzed, specifically between the energy levels E and $E + dE$,

$$N(E)dE = \frac{1}{8}\left(\frac{4\pi}{3}\right)(n+dn)^3 - \frac{1}{8}\left(\frac{4\pi}{3}\right)n^3 = \frac{1}{8}\left(\frac{4\pi}{3}\right)\left[(n+dn)^3 - n^3\right]$$

$$N(E)dE = \left(\frac{\pi}{6}\right)[n^3 + dn^3 + 3n^2dn + 3ndn^2 - n^3]$$

Due to the modest magnitude of dn, higher-order terms such as dn^2 and dn^3 may be disregarded.

$$N(E)dE = \frac{\pi}{6}3n^2dn = N(E)dE = \frac{\pi}{2}n^2dn = N(E)dE = \frac{\pi}{2n}(ndn) \tag{3.21}$$

The energy associated with an electron contained within a cubical metallic piece with sides of length 'a' is expressed as $n^2 * h^2$ (Gommes et al., 2012).

$$E = \frac{n^2h^2}{8ma^2} or; n^2 = \frac{8ma^2E}{h^2} or; n = \left[\frac{8ma^2E}{h^2}\right]^{1/2} \tag{3.22}$$

Differentiating,

$$2ndn = \frac{8ma^2dE}{h^2} or; ndn = \frac{8ma^2dE}{h^2}$$

Putting Equations 3.22 and 3.23 in 3.21,

$$N(E)dE = \frac{\pi}{2}\left(\frac{8ma^2E}{h^2}\right)^{1/2}\left[\frac{8ma^2dE}{2h^2}\right] = \frac{1}{2}\frac{\pi}{2}\left(\frac{8ma^2E}{2h^2}\right)^{1/2}\left[\frac{8ma^2dE}{h^2}\right]$$

$$= \frac{\pi}{4}\left(\frac{8ma^2}{h^2}\right)^{1/2}E^{1/2}\left(\frac{8ma^2}{h^2}\right)dE = N(E)dE = \frac{\pi}{4}\left(\frac{8ma^2}{h^2}\right)^{3/2}E^{1/2}dE \tag{3.23}$$

The Pauli exclusion principle states that no two electrons with the same spin can occupy the same quantum state simultaneously. Therefore, the number of energy levels that electrons can occupy is governed by this principle:

$$N(E)dE = 2\times\frac{\pi}{4}\left(\frac{8ma^2}{h^2}\right)^{3/2}E^{1/2}dE = \frac{\pi}{2}\left(\frac{8m}{h^2}\right)^{3/2}a^3E^{1/3}dE$$

$$= \frac{\pi}{2}(8m)^{3/2}\left(\frac{a^3}{h^3}\right)^{1/2}E^{1/2}dE = N(E)dE\frac{\pi}{2}\left(\frac{a}{h}\right)^3(8m)^{3/2}E^{1/2}dE$$

$$N(E)dE = \frac{\pi}{2}\left(\frac{a}{h}\right)^3 8(2m)^{3/2}E^{1/2}dE = \frac{4\pi}{h^3}(2m)^{3/2}a^3E^{1/2}dE$$

The density of states is defined as the quantity of energy states:

$$Z(E)dE = \frac{N(E)dE}{V}$$

$$i.e, Density\ of\ states\ Z(E)dE = \frac{\frac{4\pi}{h^3}(2m)^{3/2} a^3 E^{1/2} dE}{a^3}; [\because Volume.V = a^3]$$

$$Z(E)dE = \frac{4\pi}{h^3}(2m)^{3/2} E^{1/2} dE \quad (3.24)$$

The given expression represents the density of carriers of charge within the energy range of E to $E + dE$.

The carrier concentration, which refers to the number of electrons per unit volume within a specific energy range, is determined by multiplying the density of states $Z(E)$ by the occupancy probability $F(E)$ (Perdew and Levy, 1985).

that is, $n_c = \int Z(E)F(E)dE$

Replacing Z with E and F with E,

$$n_c = \int \frac{4\pi}{h^3}(2m)^{3/2} E^{1/2} \frac{1}{1+e^{(E-E_F)/kT}} dE$$

3.2. SEMICONDUCTING MATERIALS

A semiconducting material has a level of electrical conductivity that is significantly higher than that of an insulating material, but considerably less than that of a conducting material. The resistivity ranges from 10^{-4} to 0.5 ohms meter.

3.2.1. Properties of a Semiconductor

The following are the characteristics of a semiconductor:

1. The resistivity ranges from 10^{-4} to 0.5-ohm meter.
2. At absolute zero temperature, they exhibit insulating behavior.
3. The electrical conductivity of semiconductors is enhanced by both temperature and impurities.
4. They exhibit a negative temperature coefficient of resistivity.
5. In semiconductors, both electrons and holes serve as charge carriers and participate in conduction (Amirtharaj and Seiler, 1995).

3.2.2. Types of Semiconductors

3.2.2.1. Intrinsic Semiconductor

An intrinsic semiconductor refers to a semiconductor in their purest form. Charge carriers are solely generated through thermal agitation in this case. They have low electrical conductivity and operate at modest temperatures (Pan et al., 2019) (Figure 3.6).

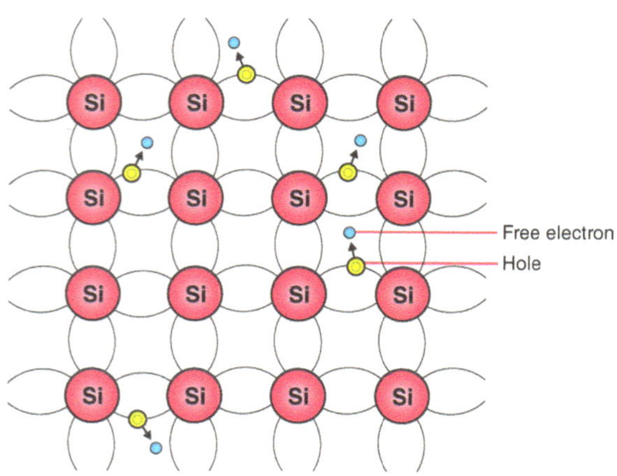

Figure 3.6. *Illustration of intrinsic semiconductors.*

Source: BYJU's; Creative Commons License.

3.2.2.2. Extrinsic Semiconductor

An extrinsic semiconductor refers to a semiconductor that has been doped with impurities. The charge carriers in this scenario are primarily produced due to these impurities and can also be formed through thermal agitation. They exhibit a substantial amount of electrical conductivity (Inkson, 1976) (Figure 3.7).

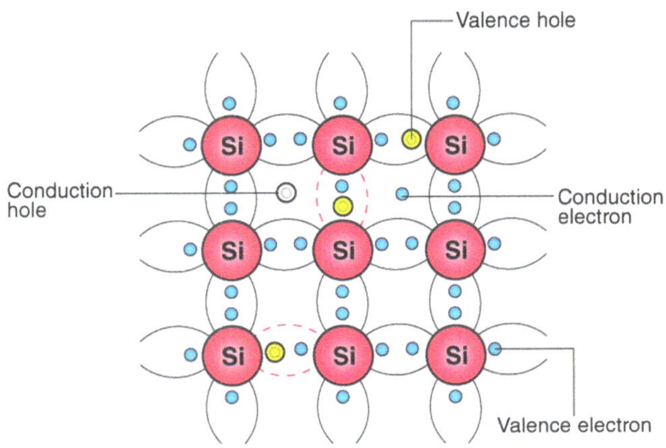

Figure 3.7. *Illustration of intrinsic semiconductors.*

Source: BYJU's; Creative Commons License.

3.2.3. Density of Electrons in the Conduction Band

The electron density in the band of conduction within the electrical energy band E and is expressed as:

$$d_n = Z(E)F(E)dE \quad (3.25)$$

$$\int_0^\infty d_n = n = \int_0^\infty Z(E)F(E)dE \quad (3.26)$$

The density associated with states in the band of conduction throughout the energy range from E to $E+dE$ is expressed as:

$$Z(E)dE = \frac{4\pi}{h^3}(2m_e^*)^{3/2}E^{1/2}dE \quad (3.27)$$

$$Z(E)dE = \frac{4\pi}{h^3}(2m_e^*)^{3/2}(E-E_c)^{1/2}dE \quad (3.28)$$

The probability of electron occupancy is determined by:

$$F(E)=\frac{1}{1+e^{(E-E_F)/kT}} \tag{3.29}$$

Substituting Equations 3.28 and 3.29 in 3.26 gives:

$$n=\frac{4\pi}{h^3}(2m_e^*)^{3/2}\int_{E_c}^{+\infty}\frac{(E-E_c)^{1/2}}{1+e^{(E-E_F)/kT}}dE \tag{3.30}$$

Because $(E - E_F)$ is larger than kT, Equation 3.30 becomes:

$$n=\frac{4\pi}{h^3}(2m_e^*)^{3/2}\int_{E_c}^{\infty}\frac{(E-E_c)^{1/2}dE}{e^{(E-E_F)/kT}},\quad n=\frac{4\pi}{h^3}(2m_e^*)^{3/2}\int_{E_c}^{+\infty}(E-E_c)^{1/2}e^{(E_F-E)kT}dE$$

$$n=\frac{4\pi}{h^3}(2m_e^*)^{3/2}e^{E_F/kT}\int_{E_c}^{\infty}(E-E_c)^{1/2}e^{-E/kT}dE \tag{3.31}$$

To find the solution to the integral in Equation 3.31, assume:

	when	when
$E - E_C = x$	$E = E_C$	$E = +\infty$
$E = E_C + x$	$E_C - E_C = x$	$+\infty - E_C$
$dE = dx$	$\therefore x = 0$	$\therefore x = +\infty$

Placing in Equation 3.31:

$$n=\frac{4\pi}{h^3}(2m_e^*)^{3/2}e^{E_F/kT}\int_0^{\infty}x^{1/2}e^{-(E_c+x)kT}dx$$

$$n=\frac{4\pi}{h^3}(2m_e^*)^{3/2}e^{(E_F-E_c)/kT}\int_0^{\infty}x^{1/2}e^{-x/kT}dx \tag{3.32}$$

The gamma function shows:

$$\int_0^{\infty}x^{1/2}e^{-x/kT}dx=\frac{(kT)^{3/2}\pi^{1/2}}{2} \tag{3.33}$$

Replacing Equations 3.33 in 3.32 (Endres et al., 2016):

$$n=\frac{4\pi}{h^3}(2m_e^*)^{3/2}e^{(E_F-E_c)/kT}\left[\frac{(kT)^{3/2}\pi^{1/2}}{2}\right]\ or\ n=2\left(\frac{2\pi m_e^*kT}{h^2}\right)^{3/2}e^{(E_F-E_e)/kT} \tag{3.34}$$

3.2.4. Density of Holes in the Valence Band

Define dp as the density of holes/volume within the valence band within the energy range from E to $E + dE$.

$$dp = Z(E)(1 - F(E))dE \quad (3.35)$$

The probability for an empty electron state, sometimes known as the presence of a hole, is denoted as 1–F (E).

$$1 - F(E) = 1 - \frac{1}{1 + e^{(E-E_F)/kT}}; 1 - F(E) = \frac{e^{(E-E_F)/kT}}{1 + e^{(E-E_F)/kT}} \quad (3.36)$$

Due to the significant difference in size between E and E_F, the quantity (E-E_F) within the valence band is negative. Thus, the term $e^{(E-EF)}$ is disregarded in the denominator.

$$\therefore 1 - F(E) = e^{(E-EF)/kT} \quad (3.37)$$

States density is:

$$Z(E)dE = \frac{4\pi}{h^3}(2m_h^*)^{3/2} E^{1/2} dE \quad (3.38)$$

In this context, m^* represents the effective mass of the hole located within the valence band. The K.E. of the hole at a level below E_v is represented by (E – E). In Equation 3.38, the word $E^{1/2}$ is substituted with (E_v – E) (Lang et al., 1981).

$$\therefore Z(E)dE = \frac{4\pi}{h^3}(2m_h^*)^{3/2}(E_v - E)^{1/2} dE \quad (3.39)$$

Substituting Equations 3.37 and 3.39 in 3.35,

$$dp = \frac{4\pi}{h^3}(2_h^*)^{3/2}(E_v - E)^{1/2} e^{(E-E_F)/kT} dE \quad (3.40)$$

The total amount of holes within the valence band across the full energy range can be determined by integrating Equation 3.40 within the specified limits $-\infty$ to E_v.

$$\int dp = p = \frac{4\pi}{h^3}(2m_h^*)^{3/2} e^{(-E_F/kT)} \int_{-\infty}^{E_v} (E_v - E)^{1/2} e^{(E/kT)} dE \quad (3.41)$$

Solving the integral:

	when	when	(3.42)
$E_v - E = x$	$E = -\infty$	$E = E_v$	
$E = -x + E_v$	$E_v + \infty = x$	$x = E_v - E_v$	(3.43)

$$\therefore dE = -dx = \infty \quad \therefore x = 0$$

Substituting in Equation 3.41, applying the gamma function:

$$\int_0^\infty x^{1/2} e^{(-x/kT)} dx = \frac{(kT)^{3/2} \pi^{1/2}}{2} \tag{3.44}$$

Replacing Equations 3.42 in 3.43 (Ley et al., 1974):

$$p = \frac{4\pi}{h^3}(2m_h^*)^{3/2} e^{(E_F - E_F)/kT} = \left[\frac{(kT)^{3/2} \pi^{1/2}}{2}\right] or\ p = 2\left(\frac{2\pi m_h^{*kT}}{h^2}\right) e^{(E_v - E_F)/kT} \tag{3.45}$$

Did you know?

Intrinsic carrier concentration is determined by the band gap energy and temperature, influencing semiconductor behavior.

3.2.5. Intrinsic Carrier Concentration

The intrinsic carrier concentration refers to the condition where the number of electrons and holes is in equilibrium. Hence, the value of n is equivalent to both the values of p and n_i.

The intrinsic carrier concentration, n_i, is mathematically expressed as the square root of the product of the electron concentration (n) and the hole concentration (p) (Sproul and Green, 1991).

$$n = \frac{4\pi}{h^3}(2m_e^*)^{3/2} e^{(E_F - E_c)/kT} \left[\frac{(kT)^{3/2} \pi^{1/2}}{2}\right] or\ n = 2\left(\frac{2\pi m_e^* kT}{h^2}\right)^{3/2} e^{(E_F - E_c)/kT}$$

$$P = \frac{4\pi}{h^3}(2m_h^*)^{3/2} e^{(E_F - E_F)/kT} \left[\frac{(kT)^{3/2} \pi^{1/2}}{2}\right] or\ p = 2\left(\frac{2\pi m_h^{*kT}}{h^2}\right)^{3/2} e^{(E_v - E_F)/kT}$$

$$n_i^2 = 2\left[\frac{2\pi kT}{3/2}\right]^{3/2} \left(m_e^* m_h^*\right)^{3/4} \frac{-Eg}{e^{2kT}} \tag{3.46}$$

- **Fermi Energy of an Intrinsic Semiconductor:** The Fermi energy level represents the critical energy level that differentiates between states that are filled with electrons and those that are empty.

At a temperature of absolute zero (0K), the Fermi energy is given by: $E_F = \frac{Ec + Ev}{2}$. The Fermi energy is precisely located between the lowest energy level of the conduction band and the highest energy level of the valence band (Su, 2008).

3.2.6. Elemental Semiconductor and Compound Semiconductor

1. **Elemental Semiconductors:** These are materials from the fourth group of the periodic table that are treated with

pentavalent or trivalent impurities to create n-type p-type conductivity.

2. **Compound Semiconductors:** These are created by combining elements from the third and fifth groups or the second and sixth groups (Fang et al., 1990).

Tip

The Hall effect is utilized in sensors and devices for measuring magnetic fields and as a diagnostic tool in material characterization.

3.2.7. Extrinsic Semiconductor

An extrinsic semiconductor is a semiconductor that has been doped with impurities. The charge carriers in this system are generated primarily by impurities but may also arise from thermal agitation. They exhibit a high level of electrical conductivity. Extrinsic semiconductors can be classified into two separate types: *N*-type semiconductors and *P*-type semiconductors (Table 3.1).

Table 3.1. *Difference between P-Type and N-Type Semiconductor*

SL. No.	N-Type	P-Type
1.	Impurity is pentavalent.	Impurity is trivalent.
2.	The majority charge carriers are electrons.	Minority carriers are holes.
3.	Donor impurity.	Acceptor impurity.
4.	The Fermi level drops as the temperature increases.	The Fermi level rises as the temperature increases.

Source: Brain Kart; Creative Commons License.

1. **Fermi Energy of Extrinsic Semiconductors at 0 K:** The term "Fermi" refers to a unit of length used in nuclear physics to measure distances on the order of 10^{-15} meters. The energy level serves as the demarcation between occupied and unoccupied states, or it represents the uppermost level at which electrons can be found.

   - For *n*-type semiconductors, it is $E_F = \frac{Ec+Ed}{2}$, that is, the Fermi energy band is precisely situated between the lowest energy level of the conduction band and the donor energy level.

   - For *p*-type, it is $E_F = \frac{Ec+Ev}{2}$, that is, it is precisely positioned between the top end of the valence band as well as the acceptor energy level (Balkan et al., 2021).

2. **Donor Energy Level:** A donor is an atom or a group of

atoms that possesses a fully occupied atomic or molecular orbital with energy higher than that of a specified reference orbital.

3. **Acceptor Energy Level:** An acceptor describes an atom or group of atoms that has an atomic or molecular orbital with lower energy than a certain standard orbital structure, and this orbital is not filled.

4. **Concentration of Holes:** In an *n*-type semiconductor, the donor energy level (*Ed*) is situated slightly below the conduction band, and *Nd* denotes the density of donor atoms per unit volume. The electron density in the conduction band is determined by:

$$n = 2\left(\frac{2\pi m_e^* kT}{h^2}\right)^{3/2} e^{(E_f - E_c)/kT} \qquad (3.47)$$

Density of an ionized donor = $N_d\ F\ (E_d)$

$$= \frac{N_d}{1 + e^{(E_d - E_f)/kT}} = \frac{1}{1 + e^{(E_d - E_f)/kT}} \qquad (3.48)$$

Because $E_d - E_F$ is larger than kT, $e^{(E\ f\ -\ Ed)/kT}$. The denominator of Equation 3.48 neglects the value of '1' due to its insignificance in comparison to the big amount.

Modifying (2) as,

$$N_d e^{(EF\ -\ Ed)/kT} \qquad (3.49)$$

During equilibrium, the density of holes in a valence band is equivalent to the density of ionized donors (Yacobi, 2003).

$$2\left(\frac{2\pi m_c^* kT}{h^2}\right)^{3/2} e^{(E_F - E_c)/kT} = N_d e^{(E_d - E_F)/kT} \qquad (3.50)$$

or,

$$\log_e\left[2\left(\frac{2\pi m_c^* kT}{h^2}\right)^{3/2}\right] + \frac{E_F - E_c}{kT} = \log_e N_d + \frac{E_d - E_F}{kT} \qquad (3.51)$$

or,

$$\frac{E_F - E_c - E_d - E_F}{kT} = +\log_e N_d - \log_e\left[2\left[\left(\frac{2\pi m_e^* kT}{h^2}\right)^{3/2}\right]\right] \qquad (3.52)$$

$$E_F = \frac{E_d + E_c}{2} + \frac{kT}{2}\log_2\left[\frac{N_d}{2\left[\frac{2\pi m_c^* kT}{h^2}\right]^{3/2}}\right] \tag{3.53}$$

Substituting E_F from Equations 3.53 in 3.47 (Gupta et al., 2016):

$$n = 2\left(\frac{2\pi m_c^* kT}{h^2}\right)^{3/2} \exp\left\{\frac{E_d - E_c}{2} + \frac{kT}{2}\log_c \frac{N_d}{\left[2\left[\frac{\frac{2\pi m_e^* kT}{h^2}}{kT}\right]\right] - E_c}\right\}$$

$$n = 2\left(\frac{2\pi m_c^* kT}{h^2}\right)^{3/2} \exp\frac{E_c + E_d - 2E_c}{2kT} + \frac{1}{2}\log_e\left[\frac{N_d}{2\left(\frac{2\pi m_c^* kT}{h^2}\right)^{3/2}}\right] \tag{3.54}$$

$$n = 2\left(\frac{2\pi m_c^* kT}{h^2}\right)^{3/2} \frac{\left(\frac{N_d}{2}\right)^{1/2}}{\left(\frac{2\pi m_c^* kT}{h^2}\right)^{3/4}} e^{(E_d - E_c)2kT}$$

$$n = (2N_d)^{1/2}\left(\frac{2\pi m_e^* kT}{h^2}\right)^{3/4} e^{(E_d - E_a)/2kT} \tag{3.55}$$

$$n = (2N_d)^{1/2}\left(\frac{2\pi m_e^* kT}{h^2}\right)^{3/4} e^{-\Delta E/2kT}$$

Where $\Delta E = E_c - E_d$ (3.56)

3.2.8. Fermi Energy of a P-Type Semiconductor

In a *p*-type semiconductor, the acceptor energy level (*Ea*) is positioned slightly above than the valence band, and *Na* represents the density of acceptor atoms per unit volume. The number of holes per unit volume in the valence band can be determined by applying the following equation:

$$p = 2\left(\frac{2\pi \pm_h^*}{h^2}\right)^{\text{ü}} e^{\text{ü}E_v - E_F \quad kT} \tag{3.57}$$

Ionized acceptors' density is:

$$\text{Na F (Ea)} = \frac{N_a}{1+e^{(E_a-E_r)kT}} = \frac{1}{1+e^{(E_a-E_r)/kT}} \tag{3.58}$$

Because Ea – EF is larger than kT, e(Ev – EF)/kT is also large. Therefore, '1' from the denominator of R.H.S. of Equation 3.58 is ignored (Gelmont et al., 1992).

Thus, Equation 3.58 is:

Na e(EF – Ea)/kT (3.59)

$$\left(\frac{n\ kT}{\quad}\right)^{3/2} e^{(E_v-E_F)/kT} \quad N\ e^{(E_F-E_a)/kT} \tag{3.60}$$

or,

$$\log_e\left[2\left(\frac{2\pi m_h^* kT}{h^2}\right)^{3/2}\right] + \frac{E_v - E_F}{kT} = \log_e N_a + \frac{E_F - E_a}{kT} \tag{3.61}$$

Rearranging,

$$\frac{E_F - E_a - E_v + E_F}{kT} - \log_e N_a + \log_e\left[2\left[\frac{2\pi_h^* kT}{h^2}\right]^{3/2}\right] \tag{3.62}$$

$$E_F = \frac{E_a + E_v}{2} - \frac{kT}{2}\log_e\left[\frac{N_a}{2\left[\frac{2\pi m_h^* kT}{h^2}\right]^{3/2}}\right] \tag{3.63}$$

At zero Kelvin, EF = $\frac{Ea + Ev}{2}$, The Fermi level is located at the midpoint between the energy level associated with the acceptor and the highest energy level within the valence band (Bocarsly et al., 1980).

3.3. HALL EFFECT

When an electric current (I) flows through a conductor, whether it is made of metal or semiconductor, and the conductor is positioned at a right angle to a magnetic field (B), it generates an electric field within the conductor (Figure 3.8).

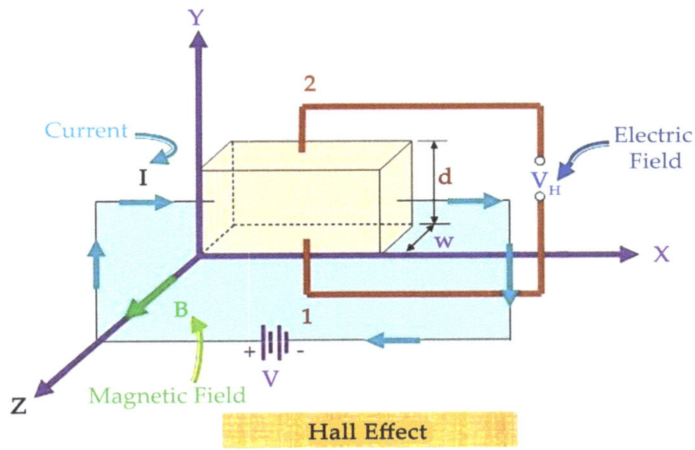

Figure 3.8. *Illustration of Hall effect.*

Source: Electronic Coach; Creative Commons License.

3.3.1. Expression of Hall Coefficient

An electric field, sometimes referred to as a potential difference, is generated in a conductor that carries a current (*I*) and is placed perpendicular to a magnetic field (*B*). The potential difference arises in a direction that is at a right angle for both the current and the magnetic field (Pichard et al., 1982).

3.3.2. Hall Effect in an N-Type Semiconductor

Consider a rectangular slab composed of a semiconducting material exhibiting n-type conductivity. The current in this material flows along the *X*-axis, whereas the magnetic field is oriented along the *Z*-axis. Therefore, the Hall voltage is developed in the *Y*-direction, as shown. The diagram clearly illustrates that the current flow is exclusively attributed to the electron movement in the *X*-direction, namely from right to left.

When an electron is in motion with a velocity v and is subjected to a magnetic field in the *Z*-direction, it will experience a downward force. Because the electrical charge for an electron is denoted by the symbol e, the downward force that the electrons experience is equal to (Qiu et al., 2020):

Bev (3.64)

The electrons are redirected in a downward direction because of this downward push (Figure 3.9).

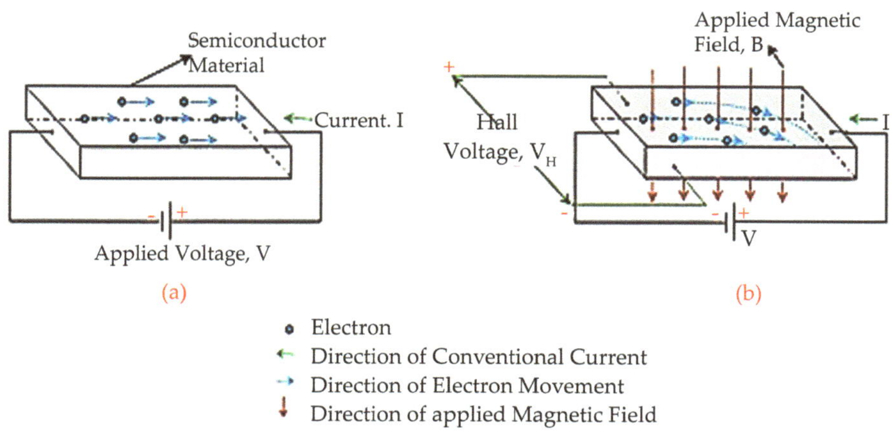

Figure 3.9. *Use of hall effect in N-type semiconductor.*

Source: Assignment Point; Creative Commons License.

In comparison to the top face, this results in the bottom face having a more unfavorable overall impression. Consequently, a potential difference in the specimen is produced between the top and the bottom. An electric field *EH*, also known as a Hall field, is created in the negative *Y*-direction because of this potential difference. A force that acts in an upward fashion on each electron is developed because of this electric field as it develops. This force is given by:

Force = eE_H (3.65)

At equilibrium, Bev balances eE_H

$\therefore$Bev = eE_H or $E_H = Bv$ (3.66)

The current density (J_x) is:

$J_x = -nev$,

$$v = \frac{-J_x}{ne} \quad (3.67)$$

Substituting Equations 3.67 in 3.66:

$$E_H = \frac{-BJ_x}{ne} \quad (3.68)$$

$$E_H = R_H J_x B \quad or, R_H = \frac{E_H}{J_x B} \tag{3.69}$$

where; $RH = -\frac{1}{ne} (for\ electrons)$; R_H is called Hall coefficient (Zarea and Ulloa, 2006).

3.3.3. Hall Coefficient in Terms of Hall Voltage

$$V_H = E_H t \tag{3.70}$$

where; t represents the thickness of a sample, while V_H represents the voltage created.

Substituting Equations 3.69 in 3.70:

$$V_H = R_H J_x Bt \tag{3.71}$$

Area (A) = (b) × (t)

$$J_x = \frac{I_x}{Area\ of\ the\ sample(A)} = \frac{I_x}{bt} \tag{3.72}$$

Substituting Equations 3.72 in 3.71:

$$V_H = \frac{R_H I_x Bt}{bt} or\ V_H = \frac{R_H I_x B}{b} \tag{3.73}$$

or, Hall coefficient is (Crescentini et al., 2021):

$$R_H = \frac{V_H b}{I_x B} \tag{3.74}$$

3.3.4. Hall Voltage

The Hall voltage arises in a conducting material an electric field and magnetic field are present, oriented at right angles to each other.

3.3.5. Experimental Setup for the Measurement of the Hall Voltage

It is shown as in Figure 3.10.

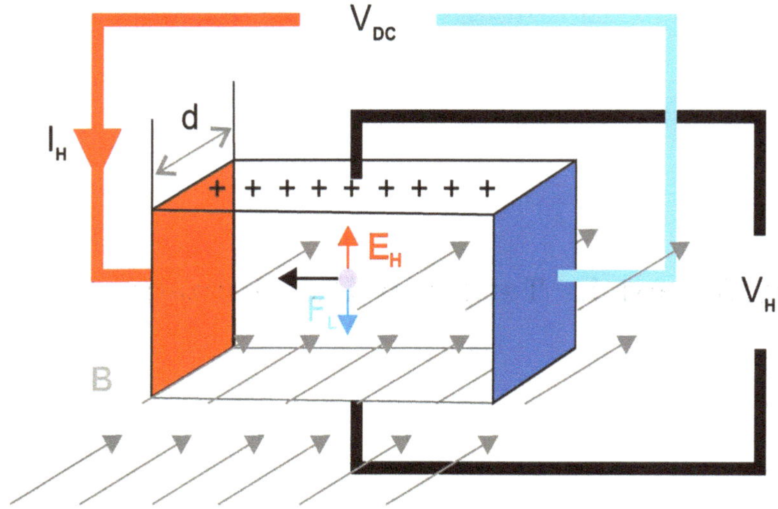

Figure 3.10. *Experimental setup for the measurement of the hall voltage.*

Source: Holmarc; Creative Commons License.

A semiconductor is taken in the shape of a rectangular slab with a thickness of '*t*' and a breadth of '*b*.' A current of I_x is then conducted through the specimen along the *X*-axis by linking it to a power source or battery (Hayashi et al., 2014).

It is positioned in such a way that *B* is applied along the *Z*-axis by positioning this sample in between two poles of an electromagnet. Hall voltage, also known as V_H, is created in the sample because of the Hall Effect.

This voltage is determined by affixing two probes to the sample in such a way that they are positioned in the middle of the lower and upper faces. Using the formula, the Hall coefficient can be calculated by measuring the Hall voltage:

$$\frac{V\ b}{I\ B}$$

The Hall coefficient enables for the determination of both carrier concentration and mobility (Castro et al., 2010).

3.3.6. Application of Hall Effect

The determination of the signs of the carriers of charge is performed. The carrier density, or concentration, is established:

$$\frac{1}{n} = \frac{1}{R_H e}$$

$$P = \frac{1}{R_H e}$$

1. Electron mobility is: $\mu_n = R_H\ e$
2. Hole mobility is: $\mu_h = R_H\ h$
3. The material's conductivity of electricity is established.

The density of magnetic flux is determined by using the previously determined Hall coefficient and the measured Hall voltage (Barlow, 1955).

3.4. MAGNETIC MATERIALS AND SUPER CONDUCTORS

Electromagnets, motors, transformers, alternators, and magnetic tapes are just a few devices that depend heavily on magnetic materials. Electronic shielding, computer core memory, and recording equipment use ferrites and metallic glasses.

1. **Bohr Magneton:** The magnetic moment resulting from the angular momentum of an electron with a quantum number (n = 1) is referred to as the Bohr magneton. It is the fundamental unit of electromagnetic moment with a specific value.

$$^{\mu}B = \frac{Eh}{4\pi m} = 9.274x10^{-24}$$

Magnetic susceptibility refers to the degree of magnetization generated in a substance relative to the strength of the magnetic field.

Magnetic permeability refers to the ability of a material to allow the passage of magnetic lines of force through it. It quantifies the extent to which it can be penetrated by a magnetic field (Makarov et al., 2022).

3.4.1. Classification of Magnetic Materials

The various types of magnetic materials are as follows:

1. Diamagnetic;
2. Ferromagnetic;
3. Paramagnetic;
4. Ferrimagnetic materials;
5. Antiferromagnetic.

3.4.1.1. Diamagnetic Materials

1. Diamagnetic materials lack persistent dipoles.
2. When an external magnetic field H is present, the motion of electrons in their orbits is changed, leading to the development of an induced magnetic moment in the atoms.

3. It has negative susceptibility and does not depend on temperature (Safarik et al., 2020) (Figure 3.11).

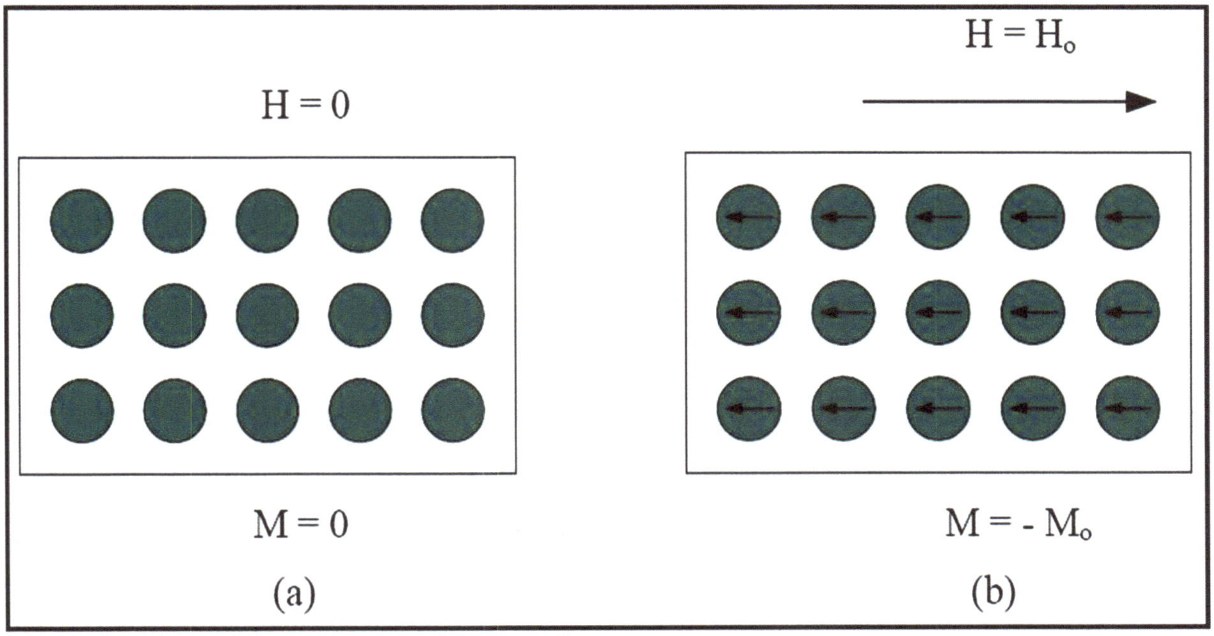

Figure 3.11. *(a) Diamagnetic material: The atoms do not possess magnetic moment, when H = 0: so M = 0. (b) When a magnetic field H o is applied, the atoms acquire induced magnetic moment in a direction opposite to the applied field that results a negative susceptibility.*

Source: Hussein S. M.; Creative Commons License.

$$N = \frac{M}{H} = -1$$

Magnetization and the induced dipoles cease to exist once the field is no longer there. Diamagnetism is an inherent characteristic shown by all substances. However, another magnetic feature prevails over it (Ueno and Iwasaka, 1994).

3.4.1.2. Para Magnetism

1. Paramagnetic materials possess permanent dipoles.
2. The dipoles have a random orientation. Thus, the overall magnetic moment is null. The dipoles exhibit a tendency to align with the external magnetic field *H*.
3. The substance undergoes magnetization. The susceptibility, denoted by ⊠, is a positive quantity that varies inversely with the temperature, *T* (Figure 3.12).

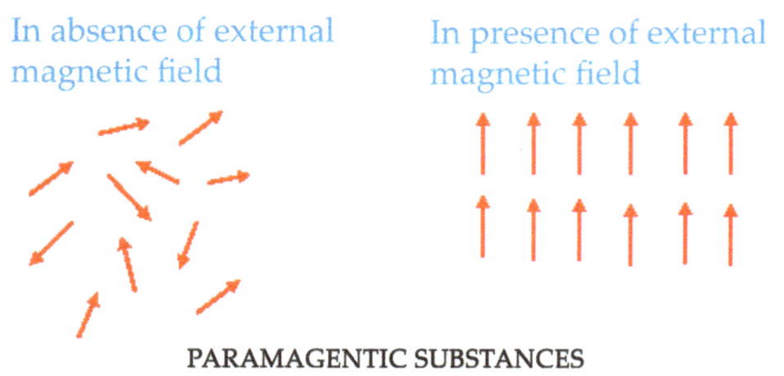

Figure 3.12. *Paramagnetism in presence and absence of magnetic field.*

Source: Toppr; Creative Commons License.

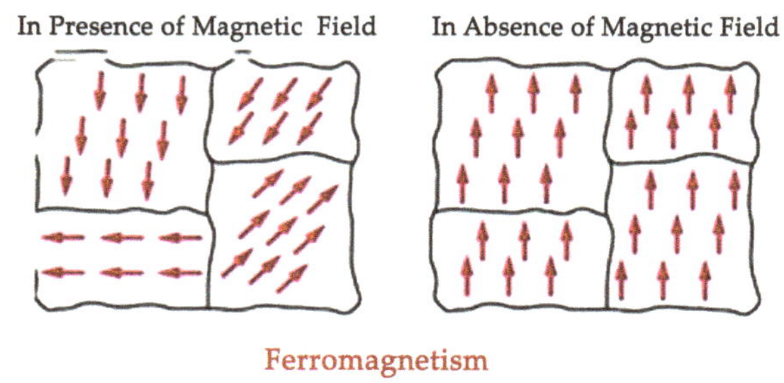

Figure 3.13. *Ferromagnetism in presence and absence of magnetic field.*

Source: Magnetism World; Creative Commons License.

$$N = \frac{c}{T} > 1$$

where; C is called Curie's constant (Abrikosov et al., 2016).

3.4.1.3. Ferromagnetism

1. Ferromagnetic materials exhibit dipoles that are permanent.
2. The dipoles are arranged in an opposing orientation throughout the domain.
3. The total magnetic moment on this domain is zero because of the intermolecular field.
4. The presence of unpaired electrons' spin magnetic moments is the cause of it. They exhibit spontaneous magnetization.
5. The material has a high susceptibility, which is temperature dependent.

$$¥ = \frac{C}{T = \theta \gg 1}$$

As the temperature T rises, the susceptibility ⊠ drops and exhibits paramagnetic characteristics above the paramagnetic Curie temperature ⊠ (Coey et al., 2010) (Figure 3.13).

3.4.2. Ferromagnetic Domain Theory

1. It is stated that a ferromagnetic substance is composed of numerous small regions of spontaneous magnetization known as domains.
2. In every domain, the magnetic moments are aligned parallel to each other.
3. The orientation of magnetism differs from one region to another, resulting in a total macroscopic magnetization of zero in an undisturbed material (Kittel, 1948).

3.4.3. Process of Domain Magnetization

1. Domain walls move in response to the application of a small magnetic field. In this scenario, the domain with magnetization aligned parallel or nearly parallel to the magnetic field increases in size, whereas the other domains decrease in size, as seen in the accompanying image.
2. By using domain rotation – when the magnetic field attains a significant magnitude, the expanding domains

become unfeasible due to the motion of domain walls (Globus, 1977) (Figure 3.14).

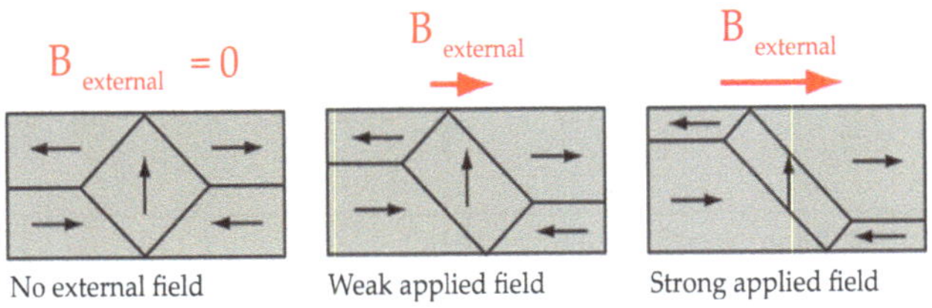

Figure 3.14. *Ferromagnetism on the basis of domain theory.*

Source: Hyper Physics; Creative Commons License.

3.4.3.1. Types of Energy Involved in the Process

Magnetostatic, exchange, and crystal anisotropy energy are involved.

3.4.4. Hysteresis loop

The *B-H* curve is plotted by representing the magnetization *H* on the *x*-axis and the magnetic induction B associated with the ferromagnetic substance on the *y*-axis (Thompson, 1909) (Figure 3.15).

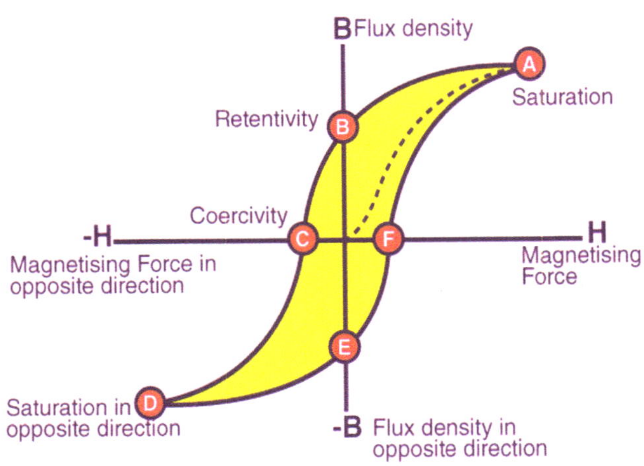

Figure 3.15. *Illustration of the hysteresis loop.*

Source: BYJU's; Creative Commons License.

The loop depicted in the picture, ABCDEA, is referred to as a Hysteresis loop.

1. In the graph, the value of *H* initially rose, causing an increase in *B* until it reached a point of saturation at *A*.
2. The value of *H* fell, followed by a fall in *B* until it reached the point of retentivity.
3. When *H* is increased in the opposite direction, *B* decreases until it reaches zero at a point called coercivity, denoted as *C*.
4. H was further augmented, causing *B* to rise in the opposite way until it achieved saturation at *D*.
5. First, the value of *H* declined. Following that, *B* also decreased till it reached the point of retentivity at *E*.
6. *H* is augmented in the opposite direction, subsequently, *B* diminishes and ultimately reaches zero.
7. When *H* was increased beyond a certain point, *B* once again attained saturation at *A*.

The loop's area represents the amount of hysteresis energy lost during the magnetization cycle of a ferromagnetic material.

1. Retentivity, also known as residual magnetism, refers to the level of magnetic induction that remains in a material at point *B* after the applied magnetic field ($H = 0$) has been removed.
2. Coercivity, also known as coercive force, refers to the minimum degree of magnetizing field *H* needed to eliminate residual magnetism from a material ($B = 0$) when applied in

the opposite direction (Sih and Jeong, 2004).

3.4.5. Types of Magnetic Materials

There are two types of magnetic materials: Soft And Hard (Table 3.2).

Table 3.2. *Soft and Hard Magnetic Materials*

Soft Magnetic Materials	Hard Magnetic Materials
They have a high susceptibility to magnetization and demagnetization	They exhibit low magnetic susceptibility and are resistant to magnetization
They possess high permeability	They possess very low permeability
Small Hysteresis loss	High Hysteresis loss
Small retentivity	Excessive retentivity
Little coercivity	Extreme coercivity
Eddy's current loss is very low	Eddy's current loss is high

Source: NIT Srinagar; Creative Commons License.

The Energy product is obtained by multiplying the retentivity and coercivity. The energy product represents the maximum amount of energy contained inside the specimen. A high energy product is crucial for the fabrication of permanent magnets. Hard magnetic materials have a high energy product (Gutfleisch et al., 2011).

Antiferromagnetism is the characteristic of certain materials in which adjacent sites exhibit two distinct types of dipoles. The electron spins of neighboring dipoles are aligned in opposing orientations. The net magnetization is zero at a temperature of 0 Kelvin. The susceptibility demonstrates a little rise when the temperature T increases initially but decreases exceeding the Neel temperature TN.

1. The susceptibility is extremely low and is in the opposite direction.
2. At first, the susceptibility exhibits a modest rise with rising temperature. At temperatures above a specific threshold called the Neel temperature, the susceptibility diminishes as the temperature increases.
3. Examples: Iron(II) oxide, Manganese(II) oxide (Fiorillo et al., 2016).

3.4.6. The Structure and Application of Ferrites

1. These are compounds formed by the combination of iron oxides with those of various other metals.
2. The magnetic moments of the sub-lattices are oriented in opposite directions and have distinct magnitudes. From a mechanical standpoint, it exhibits the properties of pure iron.
3. They exhibit significant levels of permeability and retentivity (Thakur et al., 2020).

3.4.6.1. Applications of Ferrites

1. Soft ferrites are employed for the fabrication of transformer cores.
2. Soft Ferrite bar is used to create ultrasonic sound by the mechanism of magnetostriction.
3. To manufacture magnetic film coated with ferrite, magnetic tapes, magnetic discs, and microwave drivers.
4. Hard ferrites are employed in the manufacturing of permanent magnets (Harris et al., 2009).

3.5. SUPER CONDUCTORS

When certain metals and alloys are cooled to a very low temperature known as the critical or transition temperature, their electrical resistivity abruptly becomes zero. This phenomenon is referred to as superconductivity.

The phenomenon of superconductivity was initially discovered in 1911 by the Dutch physicist H. K. Onnes while conducting an experiment to measure the electrical conductivity of metallic substances at low temperatures. Upon observation, it was noted that the resistivity of purified mercury saw a dramatic decrease to zero when cooled to a temperature of 4.2. K. Superconductivity can be observed in metallic elements as well as in metal alloys and semiconductors (Sato, 1992).

Critical temperature: The point at which a conventional conductor ceases to have electrical resistance and transforms into a superconductor is referred to as the critical temperature.

The Meissner effect describes the phenomenon where magnetic fields are forced out of the inside of a superconductor upon reaching a temperature below its critical point. The Meissner effect occurs when a superconducting material completely repels magnetic flux lines when exposed to an external magnetic field, if the temperature (T) is beneath the critical temperature (T_c) and the magnetic field strength (H) is less than the critical field strength (H_c).

In a magnetic field, a superconductor switches from superconducting to conventional conducting at the critical magnetic field. The critical field (H_c) is the magnetic field needed to remove superconductivity.

$$H_c = H_0\left[1 - \frac{T^2}{T_c^2}\right]$$

where; H is the critical field at zero Kelvin and T_c is the Transition temperature (Blatter et al., 1994) (Figure 3.16).

"Persistent current" refers to the continuous flow of electric current in a superconducting ring, maintaining its strength without any reduction, even when the magnetic field is removed, if the material remains in the superconducting state.

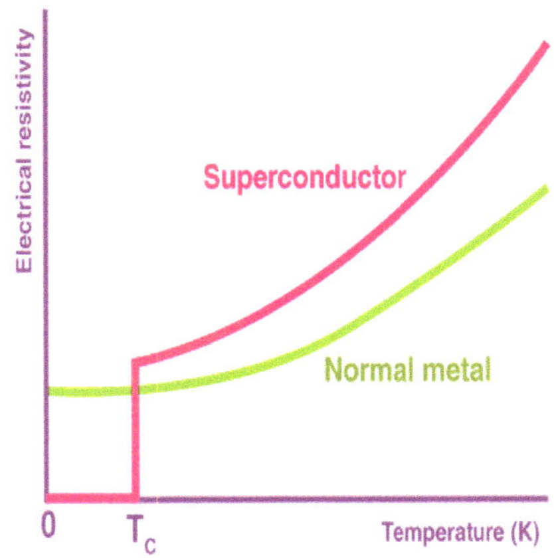

Figure 3.16. *Graph showing the temperature dependence of the electrical resistivity of normal metal and a superconductor.*

Source: BYJU's; Creative Commons License.

3.5.1. Types of Superconductors

Superconductors are of two types:

1. Type I; and
2. Type II.

A critical magnetic field causes a superconductor to abruptly switch from superconducting to conducting. These are Type I superconductors.

Type II superconductors can gradually transition from superconducting to conducting when exposed to fluctuating critical magnetic fields in an external magnetic field (Eilenberger, 1968) (Table 3.3).

Table 3.3. *Type I and Type II Superconductors*

SL. No.	Type I	Type II
1.	The Type I superconductor undergoes a sudden transition to a regular conductor when it reaches a critical magnetic field.	They rapidly lose their superconducting ability as the magnetic field increases.
2.	Only one critical field (H_c).	Two critical fields: lower (H_{c1}) and upper (Hc_2).
3.	Mixed state does not exist.	Mixed state exists.
4.	Highest value is 0.1 Tesla	Up to 30 Teslas.
5.	Examples are Al, Zn, Ga.	Examples are Vanadium and Niobium.

Source: NIT Srinagar; Creative Commons License.

3.5.2. BCS Theory

1. The BCS theory claims that superconductivity emerges when the attraction between Cooper pairs overcomes the repulsive Coulomb force.
2. A Cooper pair is a pair of electrons that are weakly coupled together through the mediation of a phonon interaction. While it may have some degree of uncertainty, this combination can be understood as follows. Imagine the motion of an electron inside a substance (Torrens and Castellano, 2018).
3. The electrostatic attraction between the positively charged nuclei and the electron of ions in the material leads to a net positive charge in the surrounding region.
4. An electron is considered "paired" when it has both opposite momentum and spin, causing it to be attracted to this force.

The BCS theory posits that electrons indeed engage in constructive interactions with atoms. It begins by assuming the presence of an attractive force between electrons, which is considered important. In conventional Type I superconductors, this force arises from the Coulombic attraction between the crystal lattice and the electron (Chen et al., 2005).

3.5.3. High T_c Super Conductor

1. High-temperature superconductors, often known as high-T_c or HTS materials, are substances that exhibit superconductivity at temperatures over 30 K (–243.2°C).
2. In the realm of superconductivity, "high temperature" is commonly defined in three ways. Technological applications are advantageous since they can utilize the higher critical temperature, which exceeds the boiling point of liquid nitrogen, as well as the greater critical magnetic field where superconductivity ceases to exist (Wendin, 1987).
3. In situations involving magnets, the elevated critical magnetic field may hold more significance than the high T_c (critical temperature) itself. Some cuprates exhibit a higher critical field of approximately 100 teslas. High-temperature superconductors possess a transition temperature (T_c) that exceeds 24 K.

For instance: $YBa_2Cu_3O_7$ has T_c = 95 K.

Characteristics of high-temperature superconductors:

1. They adhere to the resonating valence bond (RVB) theory.
2. They consist of copper oxides combined with additional elements.
3. The superconducting property of the material varies depending on the direction (Shiohara et al., 2013).

3.5.4. Magnetic Levitation

1. Superconductors actively repel magnetic fields, a phenomenon known as the Meissner effect. When a superconductor is approached by a tiny magnet, it will experience repulsion due to the generation of induced super currents that

create replicas of each pole.

2. Placing a tiny permanent magnet atop a superconductor results in levitation due to the repulsive force between them. The depicted black ceramic substance is a specimen of the superconductor composed of yttrium.
3. The suspended magnet can be sent into oscillation or rotation by hitting it with a sharp implement. The action is seen to be slowed and will stop within a few seconds (Post and Ryutov, 2000).
4. Magnetic levitating (MAGLEV) locomotives are viewed as a potential future implementation of high-temperature superconductor (HTS) advancement. To comprehend the reason for this, it is necessary to quickly examine the historical background of the railroads. The inception of trains and railways commenced in the early years of the 1800s. The contemporary standard train is not swifter (~110 mph) than the trains from the late 1890s.
5. Traditional trains are at the final stage of their development. Germany, France, and Korea have constructed high-speed trains, often known as bullet trains, that can reach speeds ranging from 150 to 180 miles per hour. This increase in velocity is attributed to enhanced railway tracks and control systems.

Nevertheless, this type of technology is also at the final stage of its development. An inhibiting element for such trains is the costly and laborious servicing of the rails. The limitation of this technique is mostly due to the physical friction between the train wheels and the metallic tracks. This issue has lead to the development of magnetically levitated trains, which operate without any friction. A brief overview of the historical progression of this development is provided.

Two fundamentally distinct ideas of magnetic suspension have emerged.

1. Electromagnetic suspension (EMS) uses electromagnets on the body of the train that are attracted to the iron rails. The vehicle magnets surround the iron guideways, generating an attracting force that lifts the train upwards.
2. The electrodynamic suspension (EDS) maintains the train in a levitated state by utilizing repulsive forces created by the currents induced in the conductive paths (El Hajjaji and Ouladsine, 2001).

3.5.5. Squids

1. The SQUID is composed of two superconductors that are separated by thin insulating layers, creating two adjacent Josephson junctions.
2. The device can be set up as a magnetometer capable of detecting extremely minute magnetic fields, including those present in live creatures.
3. Squids have been utilized to measure magnetic fields in mice brains in order to investigate whether there is sufficient magnetism to ascribe their navigational capability to an internal compass.
4. The high sensitivity of SQUID devices comes from their ability to monitor magnetic field variations with a single flux quantum. Quantization of flux in discrete units was connected to Josephson junctions.

5. The voltage measured in the SQUID device oscillates as the phase changes at the two junctions, which is influenced by changes in the magnetic flux, as long as a consistent biasing current is maintained. By quantifying the oscillations, one may assess the magnitude of the flow alteration that has taken place (Wernsdorfer, 2009).

3.5.6. Application of Superconductors

3.6.6.1. Engineering

1. It is utilized in applications such as Josephson devices, cryotrons, and SQUIDs.
2. It is utilized in computer systems, as well as in the transmission of data through lines and other mediums.

3.5.6.2. Medical

1. Superconductors are utilized in NMR imaging devices.
2. Superconducting solenoids are employed in magneto-hydrodynamic energy production to sustain plasma within the system (Tayaba et al., 2023).

3.6. DIELECTRIC MATERIALS

Insulators block electricity when voltage is applied. Materials that insulate are called dielectrics. Because electrons are firmly bound to dielectric atoms or molecules, dielectrics have no free charge carriers. Dipoles form when polar molecules align in a dielectric medium in an electric field. Polarization describes this. Electric charge storage and electrical insulation are the main uses of dielectrics.

1. They are insulators that have a band gap energy greater than 3 eV.
2. All electrons are strongly attached to atoms or molecules, and there are no existing free charges.
3. The electrons are unable to be displaced either by thermal oscillations or by regular applied voltage.
4. They are non-metallic substances with high specific resistance.
5. Examples of materials include ebonite, glass, mica, rubber, wood, and pulp (Khan and Ali, 2012).

3.7. MODERN ENGINEERING MATERIALS

3.7.1. Metallic Glasses

Metallic glasses are a type of amorphous metal that exhibit characteristics of both metal and glass. They possess strength, ductility, malleability, opacity, and brittleness. They exhibit excellent magnetic characteristics and possess a high level of resistance to corrosion. The glass transition temperature is the temperature at which the material transitions from a liquid to a solid state.

3.7.1.1. Manufacturing Metallic Glasses

The process producing metallic glasses involves the quick quenching of the melting metal alloy. This process is referred to as fast quenching (Axinte, 2012).

3.7.1.2. Melt Spinning System

1. A melt spinner has a copper roller, refractory tube, and fine nozzle.
2. Refractory tubing has an induction heater.
3. Inert gas liquefies the metallic alloy.
4. The little nozzle at the tube base expels the highly hot molten alloy.
5. Liquid alloy is poured over the fast-moving copper roller.
6. Thus, the alloy cools rapidly to form metallic glass, an amorphous solid (Chung et al., 2010).

3.7.1.3. Other Techniques

Other methods employed in the production of metallic glass ribbons are.

1. The twin roller system is a method where the molten alloy is passed through two rollers that revolve in opposite directions.
2. The melt extraction system utilizes a high-velocity roller to separate melted drops off a solid rod and shape them into a strip.

3.7.1.4. Types of Metallic Glasses

1. **Metal–Metal:** These glasses are composed of a combination of metals. For example, the elements Nickel (Ni) and Niobium (Nb) are being compared.
2. **Metal-Metalloid:** These materials are composed of a mixture of metals and metalloids. For example, the elements Fe, Co, and Ni are combined with the elements B, Si, and C (Chu et al., 2012).

3.7.1.5. Properties of Metallic Glasses

3.7.1.5.1. Structural Properties

1. The material is free from crystal imperfections including grain boundaries and dislocations.
2. Metallic glasses exhibit a tetrahedral tight packing structure.

3.7.1.5.2. Mechanical Properties

They possess exceptional strength, elasticity, and ductility.

3.7.1.5.3. Electrical Properties

1. These materials exhibit a high level of electrical resistivity, resulting in minimal eddy current loss.
2. The temperature coefficient is either zero or has a negative value.

3.7.1.5.4. Magnetic Properties

1. Metallic glasses exhibit a combination of both hard and soft magnetic characteristics.
2. These materials demonstrate a significant level of saturation magnetization, minimal losses in the core, and a tight hysteresis loop.

3.7.1.5.5. Chemical Properties

They exhibit exceptional resistance to corrosion and possess both high reactivity and stability.

3.7.1.6. Applications of Metallic Glasses

1. They function as reinforcing elements in concrete, plastic, and rubber materials.
2. They are used in the manufacturing of large flywheels for energy storage and are employed in the creation of cutting tools and other spring kinds.
3. They are used in the employment of recording device heads and as cores in high-power transformers.
4. They are used in the manufacturing of surgical instruments.
5. These materials are used as prostheses that are inserted into the human body (Nair and Priyadarshini, 2016).

3.7.2. Shape Memory Alloys

Shape-memory alloys are a type of metallic alloy that possess the ability to return back to their original size or shape when subjected to variations in temperature.

3.7.2.1. Phases of Shape Memory Alloys

1. Martensite is a phase that exists at low temperatures and is characterized by its softness and high deformability.
2. Austenite is a high-temperature phase characterized by crystalline structures and a high level of symmetry (Pons et al., 2000).

3.7.2.2. Types of Shape Memory Alloys

The two types of shape memory alloys are one-way and two-way:

1. One-way shape memory refers to a substance that demonstrates the shape memory effect exclusively when it is heated.
2. A substance that exhibits shape memory properties throughout both the heating and cooling processes is referred to as two-way shape memory.
3. Examples of these alloys include Ni-Ti alloy and Cu-Al-Ni alloy.
4. Techniques like plasma arc melting and electron beam melting are employed to fabricate shape memory alloys.

3.7.2.3. Characteristics of SMA

1. The form shape memory effect refers to the ability of a material to change its shape when subjected to low temperatures and then return to its original shape when heated. During the loading process, twinned martensite undergoes deformation and transforms into deformed martensite at low temperatures. When subjected to heat, deformed martensite changes into austenite, and when cooled, it transforms into twinned martensite (Figure 3.17).

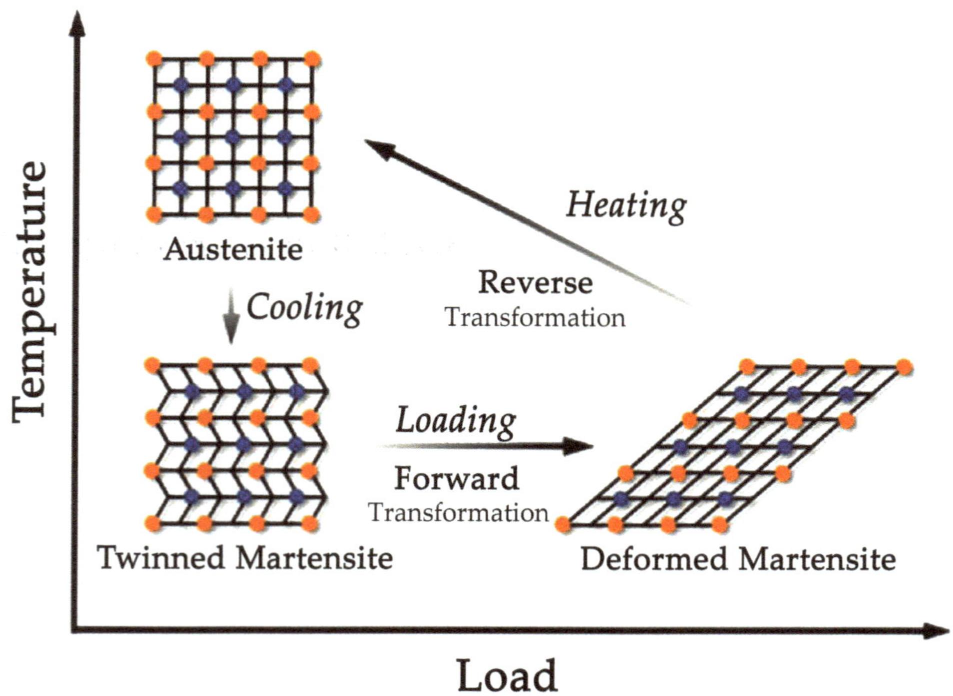

Figure 3.17. *Principle of shape memory alloy.*

Source: Kyu-Jin Cho; Creative Commons License.

2. Shape memory alloys (SMAs) undergo alterations in their electrical resistance, volume, and length as they undergo temperature-induced metamorphosis.
3. The mechanism underlying SMA is reversible.
4. Both stress as well as temperature exert a significant impact on the process of martensite transformation.
5. Pseudoelasticity refers to the behavior of a shape memory alloy (SMA) undergoes deformation when subjected to high levels of stress but regains its original shape once the load is removed. This phenomenon occurs without any change in temperature.
6. Hysteresis refers to the disparity in the transition temperature during the process of heating and cooling. The range of temperatures for the transformation of martensite to austenite during heating is slightly greater than the range for the opposite change while cooling (Lau, 2002).

3.7.2.4. Applications of Shape Memory Alloys

1. Blood clot separators and artificial hearts utilize them.
2. They are utilized in the production of toys and decorative items.
3. Eyeglass frames are constructed using these materials due to their excellent comfort and durability.
4. Metal pipes utilize them as connectors.
5. They are utilized in engines for airplanes and other settings with elevated temperatures.
6. Robotic control utilizes digital SMA stepping actuators (Song et al., 2006).

3.7.3. Nanomaterials

Nanomaterials are newly developed materials characterized by their grain size falling inside the 10^{-9} m range. They possess a multitude of grain boundaries, resulting in enhanced magnetic and chemical characteristics. They possess distinct characteristics, including immense strength, hardness, malleability, and resilience (Juan et al., 2009).

3.7.3.1. Different Forms of Nanomaterials

- Nanoparticles;
- Nano-rods and tubes;
- Nano-dots;
- Nanowires;
- Nanocomposites;
- Nanocluster;
- Nano colloids;
- Fullerenes.

3.7.3.2. Synthesis of Nanomaterials

The production of nanomaterials can be broadly categorized into two processes:

1. **Top-Down Process:** This technique involves breaking down of bulk materials into particles that are nano-sized. For example, mechanical alloying, also known as ball milling.
2. **Bottom-Up Process:** During this procedure, nanomaterials are created by assembling atoms one by one. For example: Chemical vapor deposition (Xu et al., 2013).

3.7.3.3. Preparation of Nanomaterials

There are a limited number of well-known techniques for manufacturing nanoparticles. The methods used are:

1. Ball milling;
2. Plasma arcing;
3. Chemical vapor deposition (CVD);
4. Sol-gels;
5. Laser synthesis;
6. Inert gas condensation.

3.7.3.3.1. Pulsed Laser Deposition

A high intensity laser pulse of energy is employed to vaporize carbon from graphite. Nanotubes are formed by the condensation of the evaporated carbon atoms.

1. A graphite target is enclosed within a quartz tube and placed within a muffle furnace that operates at high temperatures.
2. This quartz tube is packed with argon and heated to a temperature of 1473 K.
3. At the opposite end of the tube, a copper collector that is cooled by water is installed.
4. The graphite used as the target material contains trace amounts of nickel and cobalt, which act as catalysts to initiate the creation of nanotubes.
5. When a high-energy laser pulse strikes the target, it causes the carbon in the graphite to vaporize.
6. The carbon atoms that have evaporated are carried by the argon gas from the hotter areas to the cooler copper collector.
7. Upon reaching the cooler copper collector, the carbon atoms undergo condensation, resulting in the formation of nanotubes (Habiba et al., 2014).

3.7.3.3.2. Advantages

1. This process allows to produce single-walled carbon nanotubes with a diameter ranging from 10 to 20 nanometers.
2. The diameter of the nanotubes can be regulated by adjusting the temperature of the process.

3.7.3.3.3. Chemical Vapor Deposition

The material undergoes vaporization and is subsequently deposited onto a solid surface in a vacuum environment, resulting in the formation of nano-powder through a reaction with the substrate.

3.7.3.3.4. Description and Working

1. It involves the transfer of gas with scattered reactants over a heated surface.
2. The gas flows across the heated solid surface, delivering thermal energy.
3. It facilitates the chemical interactions of the components used in film production.
4. The byproducts are later removed.
5. Thin films with the desired qualities can be created on the surface of the substrate (Trotta and Mele, 2019).

3.7.3.3.5. Advantages

1. This technique is employed to generate nanoparticles that are free from defects.
2. The process of expanding the manufacturing capacity in the industry for mass production is

accomplished without significant challenges.

3.7.4. Properties and Applications of Nanomaterials Properties

The characteristics of nanophase materials, including chemical, electrical, conductive, and structural properties, undergo changes as the size of the particles of the material decreases.

3.7.4.1. Physical Properties

1. The distance between particles reduces as the size of metal clusters diminishes.
2. The melting point decreases as the cluster size decreases.
3. The ionization potential varies as the dimension of the nanograins changes.

3.7.4.2. Magnetic Properties

1. The bulk magnetic moment grows as the coordination number decreases.
2. Nanomaterials exhibit changes in their magnetic properties as they transition from a bulk state to a cluster state.
3. Non-magnetic materials exhibit magnetism when the size of the cluster decreases to 80 atoms (Yurkov et al., 2007).

3.7.4.3. Mechanical Properties

1. Nanophase materials exhibit low elastic strength but high plasticity.
2. The hardness and stiffness increase when the grain size decreases from 1 μm to 10 nm.
3. At low temperatures, it exhibits a high level of ductility and demonstrates superplastic behavior.

3.7.4.4. Applications of Nanophase Materials

1. Underwater nanosensor networks utilize them.
2. Nano MEMS employ them.
3. They are utilized in Molecular Nanotechnology for the purpose of constructing robotic machines known as assemblers.
4. Nanoparticles are employed in the coating of eyeglasses to provide protection against scratches and fractures.
5. Metal nanoparticles play a crucial role in the production of ionic batteries.
6. The nanoparticulate oxide of zinc is utilized in the production of very efficient sunscreen products (Siegel, 1993).

3.8. ORIGIN OF NON-LINEAR OPTICS

1. Light waves originate from electromagnetic fields.
2. When light travels through a substance, it alters the characteristics of the material.
3. The behavior is contingent upon the magnetic and electrical fields that are linked to the light.
4. Nonlinear effects were not detectable with a low-intensity conventional light beam.
5. The advent of the laser has made it feasible to generate electric fields of sufficient strength to observe intriguing nonlinear effects.
6. Therefore, when the magnetic and electrical fields reach a certain level of intensity, they will have an impact on the characteristics of the medium, thereby influencing the way the light beam travels.

3.9. CLASSIFICATION OF BIOMATERIALS

Biomaterials are divided into the following categories, according to the applications they have in the field of medicine:

1. Biomaterials made of metals and alloys;
2. Biomaterials made of ceramic;
3. Biomaterials made of polymers;
4. Composite biomaterials.

3.9.1. Metals and Alloys

Biomaterials utilize metals and alloys for their exceptional thermal and electrical conductivity, as well as their physical characteristics.

Classification of biomaterials utilizing metals and alloys:

1. Alloys composed of cobalt;
2. Titanium;
3. Stainless steel is a corrosion-resistant and stain-resistant metal alloy;
4. Prototype for a cast alloy's structure;
5. Platinum.

Metals and alloys biomaterials are utilized in the context of implantation and orthopedic procedures.

1. Surgical stainless steel is the most prevalent material used for implants.
2. Ni-Ti shape memory alloy has applications in several fields such as dentistry arch wires, tiny surgical tools, blood clot filters, and guide wires (Siraparapu et al., 2013).

3.9.2. Ceramics

Biomaterials are made from these materials for their mechanical strength and biocompatibility. Materials such as Al_2O_3 and SiO_2, quartz, porous, graphite, and alumina are discussed.

1. Tricalcium phosphate is utilized in the process of bone regeneration.

2. Ceramic implants, specifically composed of Al_2O_3 with added SiO_2 and alkali metals, are utilized in the production of femoral heads.
3. Carbon coatings are extensively used in heart valves and blood vessel transplants due to their strong compatibility with soft organs and blood.

3.9.3. Biopolymers

1. All organisms produce macromolecules spontaneously as they develop.
2. Some of the most well-known biopolymers are collagens and muco-polysaccharides like chitin.
3. One of the most common uses for collagens—which are proteins found in many kinds of animals—is in the form of solutions, gels, fibers, and tubes.
4. Drug delivery systems, blood vessels, wound dressings, and dental applications are just a few of the many medical uses for these materials (Gross and Kalra, 2002).

3.9.4. Biomaterials in Ophthalmology

1. They are utilized to enhance and sustain visual acuity.
2. In cases when the cornea, lens, and other parts of the eye are damaged or diseased, eye implants can restore their functionality.
3. The biomaterials consist of lenses, ocular shields, synthetic tears, and adjustments of corneal curvature.

3.9.5. Dental Materials

1. There are four primary categories of materials used in dental applications: polymers, composites, ceramic materials, and metal alloys (Smith, 1975).
2. Dental applications include the use of impression materials, cavity restoration, fabrication of synthetic teeth, alveolar bone restoration, and jaw support (Jones, 2001).

ACTIVITY 3.1.

Explore different types of magnetic materials (ferromagnetic, paramagnetic, etc.) and classify everyday objects based on their magnetic properties.

SUMMARY

This chapter provides an in-depth exploration of conducting materials, beginning with classical free electron theory and its applications to electrical and thermal conductivity. The Wiedemann-Franz law, Fermi-Dirac distribution function, and density of energy states are discussed to provide a deeper understanding of the behavior of free electrons in conductors.

The chapter then transitions to semiconducting materials, detailing their properties, types, and carrier concentrations. It covers intrinsic and extrinsic semiconductors, as well as the Fermi energy levels in p-type semiconductors.

Next, the Hall effect is explored including the expression of the Hall coefficient, its application in n-type semiconductors, and the experimental setup for measuring Hall voltage. Various applications of the Hall effect are also highlighted.

Magnetic materials are examined, focusing on their classification, ferromagnetic domain theory, the process of domain magnetization, and the hysteresis loop. The structure and application of ferrites are also discussed.

Superconductors are introduced, including their types, the BCS theory, high-temperature superconductors, and applications such as magnetic levitation and SQUIDs. The chapter also delves into dielectric materials and modern engineering materials like metallic glasses, shape memory alloys, and nanomaterials, emphasizing their properties and applications.

Finally, the chapter addresses the origin of non-linear optics and the classification of biomaterials, including metals, ceramics, biopolymers, and their applications in ophthalmology and dentistry. The comprehensive coverage of these topics provides a solid foundation for understanding the various aspects and applications of conducting materials in modern engineering and technology.

REVIEW QUESTIONS

1. Explain the classical free electron theory. How does it account for the electrical and thermal conductivity in metals?
2. Describe the Wiedemann-Franz law and discuss its significance in the context of conducting materials.
3. What is the Fermi-Dirac distribution function, and how does it relate to the density of energy states in a material?
4. Compare and contrast intrinsic and extrinsic semiconductors. How does the carrier concentration differ in these two types of semiconductors?
5. Discuss the Hall effect in semiconductors. How is the Hall coefficient derived, and what are its practical applications?
6. Describe the ferromagnetic domain theory. How does the process of domain magnetization contribute to the magnetic properties of materials?

MULTIPLE CHOICE QUESTIONS

1. **What does the Wiedemann-Franz law relate?**
   a. Electrical conductivity and magnetic susceptibility
   b. Thermal conductivity and specific heat
   c. Electrical conductivity and thermal conductivity
   d. Magnetic susceptibility and thermal conductivity
2. **Which of the following best describes the Fermi-Dirac distribution function?**
   a. It gives the probability of an energy state being occupied by a photon.
   b. It gives the probability of an energy state being occupied by an electron.
   c. It describes the distribution of magnetic domains in a material.
   d. It is used to calculate the thermal conductivity of metals.
3. **In semiconductors, the density of electrons in the conduction band is primarily determined by:**
   a. The temperature of the semiconductor.
   b. The magnetic field applied to the semiconductor.
   c. The number of holes in the valence band.
   d. The external voltage applied to the semiconductor.
4. **Which of the following materials is known for exhibiting superconductivity at high temperatures?**
   a. Ferrites
   b. Metallic glasses
   c. High Tc superconductors
   d. Dielectric materials
5. **The Hall effect in a semiconductor is used to measure:**
   a. The thermal conductivity of the material.
   b. The density of energy states in the material.
   c. The type and density of charge carriers.
   d. The magnetization process of the material.

Answers to Multiple Questions

1. (c); 2. (b); 3. (a); 4. (c); 5. (c)

REFERENCES

1. Abrikosov, I. A., Ponomareva, A. V., Steneteg, P., Barannikova, S. A., & Alling, B. (2016). Recent progress in simulations of the paramagnetic state of magnetic materials. *Current Opinion in Solid State and Materials Science*, *20*(2), 85–106.
2. Ahmad, S., & Mahanti, S. D. (2010). Energy and temperature dependence of relaxation time and Wiedemann-Franz law on PbTe. *Physical Review B: Condensed Matter and Materials Physics*, *81*(16), 165203.
3. Amirtharaj, P. M., & Seiler, D. G. (1995). Optical properties of semiconductors. In *Handbook of Optics* (Vol. 2, pp. 36–40).
4. Axinte, E. (2012). Metallic glasses from "alchemy" to pure science: Present and future of design, processing, and applications of glassy metals. *Materials & Design*, *35*, 518–556.
5. Balkan, N., Erol, A., Balkan, N., & Erol, A. (2021). Intrinsic and extrinsic semiconductors. In *Semiconductors for Optoelectronics: Basics and Applications* (Vol. 4, No. 1, pp. 37–78).
6. Barlow, H. E. M. (1955). The application of the Hall effect in a semiconductor to the measurement of power in an electromagnetic field. *Proceedings of the IEE-Part B: Radio and Electronic Engineering*, *102*(2), 179–185.
7. Blakemore, J. S. (1982). Approximations for Fermi-Dirac integrals, especially the function F12(⊠) used to describe electron density in a semiconductor. *Solid-State Electronics*, *25*(11), 1067–1076.
8. Blatter, G., Feigel'man, M. V., Geshkenbein, V. B., Larkin, A. I., & Vinokur, V. M. (1994). Vortices in high-temperature superconductors. *Reviews of Modern Physics*, *66*(4), 1125.
9. Bocarsly, A. B., Bookbinder, D. C., Dominey, R. N., Lewis, N. S., & Wrighton, M. S. (1980). Photoreduction at illuminated p-type semiconducting silicon photoelectrodes: Evidence for Fermi level pinning. *Journal of the American Chemical Society*, *102*(11), 3683–3688.
10. Burger, N., Laachachi, A., Ferriol, M., Lutz, M., Toniazzo, V., & Ruch, D. (2016). Review of thermal conductivity in composites: Mechanisms, parameters, and theory. *Progress in Polymer Science*, *61*, 1–28.
11. Cahill, D. G. (1990). Thermal conductivity measurement from 30 to 750 K: The 3⊠ method. *Review of Scientific Instruments*, *61*(2), 802–808.
12. Carruthers, P. (1961). Theory of thermal conductivity of solids at low temperatures. *Reviews of Modern Physics*, *33*(1), 92.
13. Castro, H., Galvis, J., & Castro, S. (2010). Automated setup for Van der Pauw Hall measurements. *IEEE Transactions on Instrumentation and Measurement*, *60*(1), 198–205.
14. Chen, Q., Stajic, J., Tan, S., & Levin, K. (2005). BCS–BEC crossover: From high-temperature superconductors to ultracold superfluids. *Physics Reports*, *412*(1), 1–88.
15. Chu, J. P., Jang, J. S. C., Huang, J. C., Chou, H. S., Yang, Y., Ye, J. C., & Rullyani,

C. (2012). Thin film metallic glasses: Unique properties and potential applications. *Thin Solid Films*, *520*(16), 5097–5122.

16. Chung, S., Ingle, N. P., Montero, G. A., Kim, S. H., & King, M. W. (2010). Bioresorbable elastomeric vascular tissue engineering scaffolds via melt spinning and electrospinning. *Acta Biomaterialia*, *6*(6), 1958–1967.
17. Coey, J. M. D., Stamenov, P., Gunning, R. D., Venkatesan, M., & Paul, K. (2010). Ferromagnetism in defect-ridden oxides and related materials. *New Journal of Physics*, *12*(5), 53025.
18. Crescentini, M., Syeda, S. F., & Gibiino, G. P. (2021). Hall-effect current sensors: Principles of operation and implementation techniques. *IEEE Sensors Journal*, *22*(11), 10137–10151.
19. Dubey, N., & Leclerc, M. (2011). Conducting polymers: Efficient thermoelectric materials. *Journal of Polymer Science Part B: Polymer Physics*, *49*(7), 467–475.
20. Duboz, J. Y., Costard, E., Rosencher, E., Bois, P., Nagle, J., Berset, J. M., & Ortega, J. M. (1995). Electron relaxation time measurements in GaAs/AlGaAs quantum wells: Intersubband absorption saturation by a free-electron laser. *Journal of Applied Physics*, *77*(12), 6492–6495.
21. Eilenberger, G. (1968). Transformation of Gorkov's equation for type II superconductors into transport-like equations. *Zeitschrift für Physik A: Hadrons and Nuclei*, *214*, 195–213.
22. El Hajjaji, A., & Ouladsine, M. (2001). Modeling and nonlinear control of magnetic levitation systems. *IEEE Transactions on Industrial Electronics*, *48*(4), 831–838.
23. Endres, J., Egger, D. A., Kulbak, M., Kerner, R. A., Zhao, L., Silver, S. H., & Kahn, A. (2016). Valence and conduction band densities of states of metal halide perovskites: A combined experimental–theoretical study. *The Journal of Physical Chemistry Letters*, *7*(14), 2722–2729.
24. Fang, S. F., Adomi, K., Iyer, S., Morkoc, H., Zabel, H., Choi, C., & Otsuka, N. (1990). Gallium arsenide and other compound semiconductors on silicon. *Journal of Applied Physics*, *68*(7), 31–58.
25. Fiorillo, F., Bertotti, G., Appino, C., & Pasquale, M. (2016). Soft magnetic materials. In *Wiley Encyclopedia of Electrical and Electronics Engineering* (Vol. 1, pp. 1–42). John Wiley & Sons, Inc.
26. Friedman, S. P. (2005). Soil properties influencing apparent electrical conductivity: A review. *Computers and Electronics in Agriculture*, *46*(1–3), 45–70.
27. Gelmont, B., Shur, M., & Moglesture, C. (1992). Theory of junction between two-dimensional electron gas and p-type semiconductor. *IEEE Transactions on Electron Devices*, *39*(5), 1216–1222.
28. Globus, A. (1977). I. Magnetization mechanisms: Some physical considerations about the domain wall size theory of magnetization mechanisms. *Le Journal de Physique Colloques*, *38*(1), 1–1.
29. Gommes, C. J., Jiao, Y., & Torquato, S. (2012). Density of states for a specified

correlation function and the energy landscape. *Physical Review Letters*, *108*(8), 080601.

30. Gross, R. A., & Kalra, B. (2002). Biodegradable polymers for the environment. *Science*, *297*(5582), 803–807.
31. Gupta, K. M., Gupta, N., Gupta, K. M., & Gupta, N. (2016). Carrier transport in semiconductors. *Advanced Semiconducting Materials and Devices*, *2*(3), 87–144.
32. Gutfleisch, O., Willard, M. A., Brück, E., Chen, C. H., Sankar, S. G., & Liu, J. P. (2011). Magnetic materials and devices for the 21st century: Stronger, lighter, and more energy efficient. *Advanced Materials*, *23*(7), 821–842.
33. Habiba, K., Makarov, V. I., Weiner, B. R., & Morell, G. (2014). Fabrication of nanomaterials by pulsed laser synthesis. In *Manufacturing Nanostructures* (pp. 263–292).
34. Harris, V. G., Geiler, A., Chen, Y., Yoon, S. D., Wu, M., Yang, A., & Vittoria, C. (2009). Recent advances in processing and applications of microwave ferrites. *Journal of Magnetism and Magnetic Materials*, *321*(14), 2035–2047.
35. Hayashi, M., Kim, J., Yamanouchi, M., & Ohno, H. (2014). Quantitative characterization of the spin-orbit torque using harmonic Hall voltage measurements. *Physical Review B*, *89*(14), 144425.
36. Inkson, J. C. (1976). The effect of electron interaction on the band gap of extrinsic semiconductors. *Journal of Physics C: Solid State Physics*, *9*(7), 1177.
37. Jones, F. H. (2001). Teeth and bones: Applications of surface science to dental materials and related biomaterials. *Surface Science Reports*, *42*(3–5), 75–205.
38. Juan, J. S., Nó, M. L., & Schuh, C. A. (2009). Nanoscale shape-memory alloys for ultrahigh mechanical damping. *Nature Nanotechnology*, *4*(7), 415–419.
39. Justice, P., Marshman, E., & Singh, C. (2019). Student understanding of Fermi energy, the Fermi–Dirac distribution and total electronic energy of a free electron gas. *European Journal of Physics*, *41*(1), 015704.
40. Kerner, E. H. (1956). The electrical conductivity of composite media. *Proceedings of the Physical Society. Section B*, *69*(8), 802.
41. Khan, M. T., & Ali, S. M. (2012). A brief review of measuring techniques for characterization of dielectric materials. *International Journal of Information Technology and Electrical Engineering*, *1*(1), 1–5.
42. Kittel, C. (1948). Domain theory and the dependence of the coercive force of fine ferromagnetic powders on particle size. *Physical Review*, *73*(7), 810.
43. Lang, J. K., Baer, Y., & Cox, P. A. (1981). Study of the 4f and valence band density of states in rare-earth metals. II. Experiment and results. *Journal of Physics F: Metal Physics*, *11*(1), 121.
44. Lau, K. T. (2002). Vibration characteristics of SMA composite beams with different boundary conditions. *Materials & Design*, *23*(8), 741–749.
45. Ley, L., Pollak, R. A., McFeely, F. R., Kowalczyk, S. P., & Shirley, D. A. (1974). Total valence-band densities of states of III-V and II-VI compounds from x-ray photoemission spectroscopy. *Physical Review B*, *9*(2), 600.

46. Li, D. L., & Yu, E. L. (2011). An approach based on the classical free-electron theory to study the electroplastic effect. *Advanced Materials Research*, *148*, 71–74.

47. Luong, M., & Shaw, A. W. (1971). Quantum transport theory of impurity-scattering-limited mobility in n-type semiconductors including electron-electron scattering. *Physical Review B*, *4*(8), 2436.

48. Makarov, D., Volkov, O. M., Kákay, A., Pylypovskyi, O. V., Budinská, B., & Dobrovolskiy, O. V. (2022). New dimension in magnetism and superconductivity: 3D and curvilinear nanoarchitectures. *Advanced Materials*, *34*(3), 2101758.

49. McDougall, J., & Stoner, E. C. (1938). The computation of Fermi-Dirac functions. *Philosophical Transactions of the Royal Society of London. Series A, Mathematical and Physical Sciences*, *237*(773), 67–104.

50. Mills, R. L. (2001). The nature of free electrons in superfluid helium—A test of quantum mechanics and a basis to review its foundations and make a comparison to classical theory. *International Journal of Hydrogen Energy, 26*(10), 1059–1096.

51. Nair, B., & Priyadarshini, B. G. (2016). Process, structure, property, and applications of metallic glasses. *AIMS Materials Science, 3*(3), 1022–1053.

52. Pal, T., Banerjee, S., Manna, P. K., & Kar, K. K. (2020). Characteristics of conducting polymers. In *Handbook of Nanocomposite Supercapacitor Materials I: Characteristics* (pp. 247–268).

53. Pan, Z., Röhr, J. A., Ye, Z., Fishman, Z. S., Zhu, Q., Shen, X., & Hu, S. (2019). Elucidating charge separation in particulate photocatalysts using nearly intrinsic semiconductors with small asymmetric band bending. *Sustainable Energy & Fuels, 3*(3), 850–864.

54. Park, S. J., & Donnet, J. B. (1998). Evaluation of the distribution function of adsorption site energies based on the Fermi–Dirac's law in a monolayer. *Journal of Colloid and Interface Science, 200*(1), 46–51.

55. Perdew, J. P., & Levy, M. (1985). Extrema of the density functional for the energy: Excited states from the ground-state theory. *Physical Review B, 31*(10), 6264.

56. Pichard, C. R., Tosser, A. J., & Tellier, C. R. (1982). General expression for the Hall coefficient of thin metal films. *Journal of Materials Science Letters, 1*(6), 260–263.

57. Pons, J., Chernenko, V. A., Santamarta, R., & Cesari, E. (2000). Crystal structure of martensitic phases in Ni–Mn–Ga shape memory alloys. *Acta Materialia, 48*(12), 3027–3038.

58. Post, R. F., & Ryutov, D. D. (2000). The Inductrack: A simpler approach to magnetic levitation. *IEEE Transactions on Applied Superconductivity, 10*(1), 901–904.

59. Qiu, G., Niu, C., Wang, Y., Si, M., Zhang, Z., Wu, W., & Ye, P. D. (2020). Quantum Hall effect of Weyl fermions in n-type semiconducting tellurene. *Nature Nanotechnology, 15*(7), 585–591.

60. Radzuan, N. A. M., Sulong, A. B., & Sahari, J. (2017). A review of electrical conductivity models for conductive polymer composite. *International Journal of Hydrogen Energy, 42*(14), 9262–9273.

61. Safarik, I., Prochazkova, J., Baldikova, E., Timko, M., Kopcansky, P., Rajnak, M., &

Pospiskova, K. (2020). Modification of diamagnetic materials using magnetic fluids. *Ukrainian Journal of Physics, 65*(9), 751–751.

62. Sarachik, E. S., & Schappert, G. T. (1970). Classical theory of the scattering of intense laser radiation by free electrons. *Physical Review D, 1*(10), 2738.
63. Sato, K. I. (1992). Superconductors. *Physics World, 5*(7), 37.
64. Shiohara, Y., Yoshizumi, M., Takagi, Y., & Izumi, T. (2013). Future prospects of high Tc superconductors-coated conductors and their applications. *Physica C: Superconductivity, 484,* 1–5.
65. Siegel, R. W. (1993). Nanophase materials assembled from atom clusters. *Materials Science and Engineering: B, 19*(1, 2), 37–43.
66. Sih, G. C., & Jeong, D. Y. (2004). Hysteresis loops predicted by isoenergy density theory for polycrystals. Part I: Fundamentals of non-equilibrium thermal–mechanical coupling effects. *Theoretical and Applied Fracture Mechanics, 41*(1–3), 233–266.
67. Siraparapu, Y. D., Bassa, S., & Sanasi, P. D. (2013). A review on recent applications of biomaterials. *International Journal of Scientific Research, 1,* 70–75.
68. Smith, D. C. (1975). Biomaterials in dentistry. *Journal of Dental Research, 54*(2_ suppl), 145–152.
69. Song, G., Ma, N., & Li, H. N. (2006). Applications of shape memory alloys in civil structures. *Engineering Structures, 28*(9), 1266–1274.
70. Sproul, A. B., & Green, M. A. (1991). Improved value for the silicon intrinsic carrier concentration from 275 to 375 K. *Journal of Applied Physics, 70*(2), 846–854.
71. Su, C. H. (2008). Energy band gap, intrinsic carrier concentration, and Fermi level of CdTe bulk crystal between 304 and 1067 K. *Journal of Applied Physics, 103*(8), 52–55.
72. Tallman, D. E., Spinks, G., Dominis, A., & Wallace, G. G. (2002). Electroactive conducting polymers for corrosion control: Part 1. General introduction and a review of non-ferrous metals. *Journal of Solid State Electrochemistry, 6,* 73–84.
73. Tayaba, S., Sethi, H., Shahid, H., Malik, R., Ikram, M., Ali, S., & Maqbool, M. (2023). Silicon-germanium and carbon-based superconductors for electronic, industrial, and medical applications. *Materials Science and Engineering: B, 290,* 116332.
74. Thakur, P., Chahar, D., Taneja, S., Bhalla, N., & Thakur, A. (2020). A review on MnZn ferrites: Synthesis, characterization, and applications. *Ceramics International, 46*(10), 15740–15763.
75. Thompson, S. P. (1909). On hysteresis loops and Lissajous' figures, and on the energy wasted in a hysteresis loop. *Proceedings of the Physical Society of London, 22*(1), 454.
76. Torrens, F., & Castellano, G. (2018). Superconductors, superconductivity, BCS theory, and entangled photons for quantum computing. In *Physical Chemistry for Engineering and Applied Sciences* (Vol. 2, pp. 401–410). Apple Academic Press.
77. Trotta, F., & Mele, A. (2019). Nanomaterials: Classification and properties. *Nanosponges: Synthesis and Applications, 5*(2), 1–26.

78. Ueno, S., & Iwasaka, M. (1994). Properties of diamagnetic fluid in high gradient magnetic fields. *Journal of Applied Physics, 75*(10), 7177–7179.

79. Völklein, F., Reith, H., Cornelius, T. W., Rauber, M., & Neumann, R. (2009). The experimental investigation of thermal conductivity and the Wiedemann–Franz law for single metallic nanowires. *Nanotechnology, 20*(32), 325706.

80. Wan, M. (2008). Introduction of conducting polymers. *Conducting Polymers with Micro or Nanometer Structure, 4*(1), 1–15.

81. Wendin, G. (1987). High-energy spectroscopies and high-Tc superconductors: An overview. *Le Journal de Physique Colloques, 48*(9), 1155–1157.

82. Wernsdorfer, W. (2009). From micro-to nano-SQUIDs: Applications to nanomagnetism. *Superconductor Science and Technology, 22*(6), 064013.

83. Xu, H., Zeiger, B. W., & Suslick, K. S. (2013). Sonochemical synthesis of nanomaterials. *Chemical Society Reviews, 42*(7), 2555–2567.

84. Yacobi, B. G. (2003). Basic properties of semiconductors. In *Semiconductor Materials: An Introduction to Basic Principles* (Vol. 1, pp. 59–105).

85. Yurkov, G. Y., Fionov, A. S., Koksharov, Y. A., Koleso, V. V., & Gubin, S. P. (2007). Electrical and magnetic properties of nanomaterials containing iron or cobalt nanoparticles. *Inorganic Materials, 43,* 834–844.

86. Zarea, M., & Ulloa, S. E. (2006). Spin Hall effect in two-dimensional p-type semiconductors in a magnetic field. *Physical Review B—Condensed Matter and Materials Physics, 73*(16), 165306.

CHAPTER

4 Electromagnetism and Electrodynamics

LEARNING OBJECTIVES

After studying this chapter, you will be able to:

- Understand the evolution of electromagnetic theory and its fundamental concepts.
- Explain Maxwell's equations, including Gauss's Law for electricity and magnetism, Faraday's Law of Induction, and Ampère's Law with Maxwell's addition.
- Describe the properties and behavior of electromagnetic waves, including the speed of light in a vacuum.
- Analyze the concepts of energy and momentum in electromagnetic waves, focusing on the Poynting vector, radiation pressure, and wave intensity.
- Understand the principles of electromagnetic induction, including inductance, self-inductance, and mutual inductance.
- Explore the application of electromagnetic waves in communication technologies, such as radio waves, microwaves, and fiber optics.
- Examine the role of electromagnetic induction in power generation, including electric generators, AC and DC systems, and grid distribution.

KEY TERMS FROM THIS CHAPTER

Ampère's law with maxwell's addition
Faraday's law of induction
Gauss's law for magnetism
Mutual inductance
Electromagnetic induction
Gauss's law for electricity
Maxwell's equations

UNIT INTRODUCTION

Electromagnetism, as one of the four fundamental forces of nature, plays a crucial role in understanding a range of scientific phenomena and current technological advancements. This chapter presents the fundamental concepts and equations that govern the behavior of magnetic and electric fields, as well as their interactions. It will begin by examining the chronological progression of electromagnetic theory, and will subsequently delve into the basic principles that are crucial for grasping more intricate subjects (Brevik, 1979).

4.1. ELECTROMAGNETIC THEORY

The field of electromagnetic theory has seen substantial development since the first discoveries regarding electric and magnetic processes. Both Michael Faraday and James Clerk Maxwell were influential in the development of the fundamental ideas of this discipline (Rayleigh, 1881).

4.1.1. Evolution of Electromagnetic Theory

Faraday's investigations in the first decade of the nineteenth century established the correlation between magnetism and electricity. His discovery of electromagnetic induction demonstrated that a fluctuating magnetic field could generate electrical current in a conductor. The mathematical expression of Faraday's Law of Induction is as follows:

$$\varepsilon = \frac{-d\phi_\beta}{d_t}$$

Where ε is the induced electromotive force (also known as EMF) and ϕ_β is the magnetic flux (Kiehn, 1999).

James Clerk Maxwell expanded upon Faraday's research and developed the classical theory of electromagnetism. During the 1860s, Maxwell consolidated the preexisting principles of electricity and magnetism into a set of four partial differential equations, which are now recognized as Maxwell's Equations. These equations delineate the generation and modification of electric and magnetic fields through their interactions with each other, as well as with charges and currents:

- Gauss's Law for Electricity:

$$\nabla \cdot E = \frac{\rho}{\varepsilon_0}$$

- Gauss's Law for Magnetism:

$$\nabla . B = 0$$

- Faraday's Law of Induction:

$$\nabla \times E = \frac{-\partial \beta}{\partial t}$$

- Ampère's Law (with Maxwell's Addition):

$$\nabla \times B = \mu_0 J + \mu_0 \varepsilon_0 \frac{\partial E}{\partial t}$$

Maxwell's unification established that magnetic and electrical fields propagate as waves in space, traveling at the speed of light. This discovery substantially connects electromagnetism with optics (Hayakawa et al., 2004).

4.1.2. Fundamental Concepts

To comprehend electromagnetism, one must master the following essential concepts:

4.1.2.1. Electric Fields (E)

It is a vector field that envelops a charged particle and signifies the exertion of force on other charges situated inside the field. It is given by (Hajnal, 1987) (Figure 4.1):

$$E = \frac{F}{q}$$

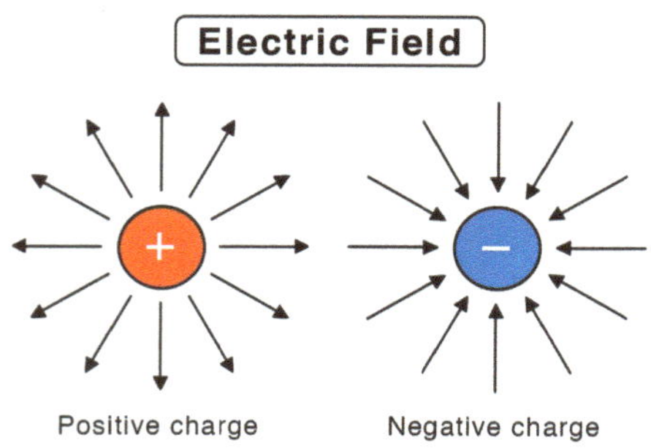

Figure 4.1. *Electric field around a positive and negative charge.*

Source: Science Facts, Creative Commons License.

4.1.2.2. Magnetic Fields)

Traveling electric charges and moving magnetic dipoles create a magnetic field, which then exerts a force on other charged particles and magnetic dipoles. The magnetic force acting on a moving charge can be determined using the following equation (Hartree, 1931) (Figure 4.2):

$$F = q(v \times B)$$

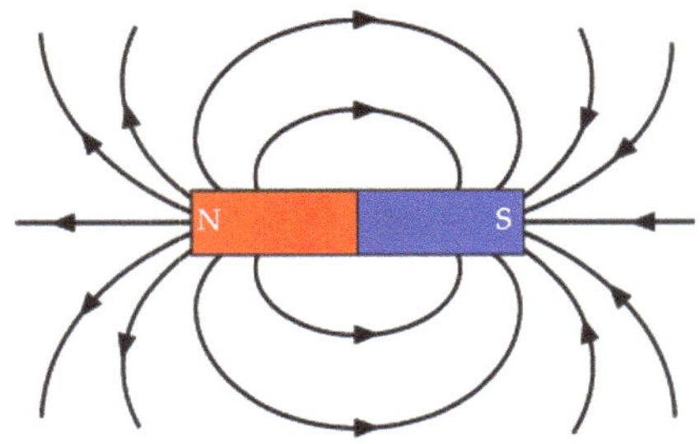

Figure 4.2. *Illustration of magnetic field around a magnet.*

Source: BBC, Creative Commons License.

4.1.2.3. Charge and Current

Charge (q) is an intrinsic characteristic of matter that causes a force to be exerted on it when it is exposed to an electric or magnetic field. Electric current (I) is the flow of electric charge and is precisely defined as (Shaarawi, 2000):

$$I = \frac{dQ}{dt}$$

where (Q) is the charge passing through a point in each time (t).

4.1.2.4. Magnetic Flux (Φ_B)

It is calculated as:

$$\Phi_B = \int B \cdot dA$$

where (dA) is a vector that represents a very small region on the surface.

These basic principles serve as the foundation for comprehending more intricate phenomena in electromagnetics and electrodynamics, as examined in the following sections (Körpınar et al., 2022).

4.2. MAXWELL'S EQUATIONS

4.2.1. Gauss's Law for Electricity

Gauss's Law for Electricity, a component of Maxwell's equations, is a fundamental concept that describes the relationship between electric charge and the resultant electric field. The assertion posits that the electric flux across a closed surface is directly proportional to the amount of charge enclosed by this surface. One may comprehend and derive this rule by using both integral and differential forms (Figure 4.3).

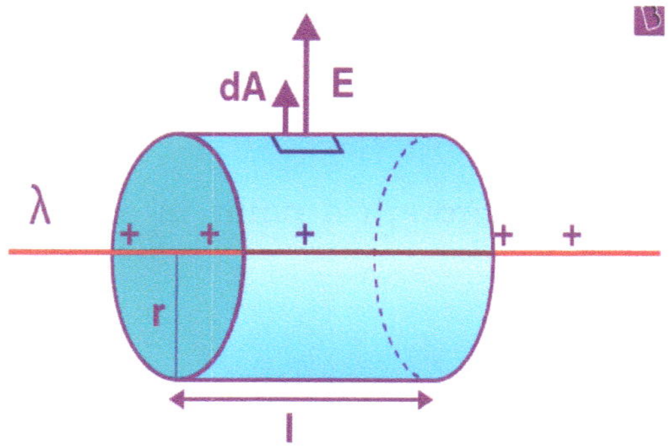

Figure 4.3. *Electric field due to infinite wire.*

Source: BYJU's, Creative Commons License.

The mathematical expression for Gauss's Law for electricity is given by:

$$\oint_{surface} E \times dA = \frac{Q_{enc}}{\epsilon_0}$$

where, $\left(\oint_{surface} E \cdot dA\right)$ is a closed surface electric flux; (dA) is the area vector; (Q_{enc}) is the total enclosed charge; and (ϵ_0) is the free space permittivity (Maries et al., 2017).

4.2.1.1. Integral Form of Gauss's Law

Gauss's Law, stated in integral form, asserts that the total electric flux going through a closed surface is equal to the amount of charge enclosed by the surface divided by the permittivity of free space. This form is useful for calculating the electric field of charge distributions that exhibit symmetry.

4.2.1.1.1. Derivation of the Integral Form:

Consider a charge (q) is located precisely in the center of a spherical surface (Christodoulou & Kazanas, 2019).

Coulomb's law says that (E) at a distance (r) from (q) placed at the center of a sphere with a radius may be determined:

$$E = \frac{q}{4\pi\epsilon_0 r^2}\hat{r}$$

where ($\hat{r}$) represents the unit vector pointing in the radial direction.

Determine the electric flow (Φ_E) over the spherical surface by performing the necessary calculations.

The electric flux passing through a tiny area element (dA) on the surface of a sphere equals:

$$d\Phi_E = E \cdot dA$$

Since (E) and (dA) are parallel and (E) is constant over the surface,

$$d\Phi_E = E\,dA$$

The total flux is:

$$\Phi_E = \oint_{surface} E \cdot dA = \oint_{surface} E\,dA = E\oint_{surface} dA$$

The surface area of a sphere is $(4\pi r^2)$:

$$\Phi_E = E \cdot 4\pi r^2$$

Substitute the electric field (E):

$$\Phi_E = \left(\frac{q}{4\pi\epsilon_0 r^2}\right) \cdot 4\pi r^2 = \frac{q}{\epsilon_0}$$

Therefore, Gauss's Law in integral form is (Kaplan & Stryker, 2020):

$$\oint_{surface} E \cdot dA = \frac{Q_{enc}}{\epsilon_0}$$

4.2.1.2. Differential Form of Gauss's Law

Gauss's Law in differential form relates the divergence of the electric field to the charge density at a certain place. It is represented as:

$$\nabla \cdot E = \frac{\rho}{\epsilon_0}$$

where, $(\nabla \cdot E)$ is electric field divergence; and (ρ) is the charge density (Sanders et al., 2014).

4.2.1.2.1. Derivation of Differential Form

- From Gauss's Law:

$$\oint_{surface} E \cdot dA = \frac{Q_{enc}}{\epsilon_0}$$

1. **Apply the Divergence Theorem:** The divergence theorem states that the flux of a vector field through a closed surface is equal to the volume integral of the divergence of the field within the surface:

$$\oint_{surface} E \cdot dA = \int_{volume} (\nabla \cdot E)\,dV$$

2. **Relate the Enclosed Charge to the Volume Charge Density:** The enclosed charge (Q_{enc}) can be expressed as the volume integral of (ρ):

$$Q_{enc} = \int_{volume} \rho dV$$

3. Combine the Equations:

$$\int_{volume} (\nabla \cdot E) dV = \frac{1}{\epsilon_0} \int_{volume} \rho dV$$

4. Since the Volume Integrals Are Equal for Any Arbitrary Volume:

$$\nabla \cdot E = \frac{\rho}{\epsilon_0}$$

Both the differential and integral formulations of Gauss's Law are crucial for resolving electrostatic issues and comprehending the behavior of electric fields in diverse arrangements (Warnick et al., 1996).

4.2.2. Gauss's Law for Magnetism

This is a key equation in Maxwell's equations that is crucial for understanding the characteristics of magnetic fields. The principle asserts that the total magnetic flux through a closed surface is always zero, implying the absence of magnetic monopoles. Consequently, magnetic field lines form closed-loop configuration and do not have any starting or ending points.

The mathematical expression is as follows:

$$\oint_{surface} B \cdot dA = 0$$

where, represents the magnetic flux passing through a closed surface; and (dA) is the differential area vector on the closed surface (Poole, 2018).

4.2.2.1. Integral Form of Gauss's Law for Magnetism

Gauss's Law, in integral form, states that the net magnetic flux passing through a closed surface is zero. This form is especially valuable for comprehending the lack of magnetic monopoles and the unbroken nature of magnetic field lines.

4.2.2.1.1. Derivation of the Integral Form

1. **Imagine a Surface That Surrounds a Certain Volume (Reinhardt, 1997):** To determine the total magnetic flux going through a closed surface, it is necessary to calculate the net value. The closed loop of magnetic field lines necessitates that the number of field lines entering a surface is equal to the number of field lines departing the surface.
2. **Determine the Magnetic Flux Over the Enclosed Surface:** The magnetic flux passing through a small area element on the closed surface:

$$d\Phi_B = B \cdot dA$$

The total flux:

$$\Phi_B = \oint_{surface} B \cdot dA$$

3. **Calculate the Net Flux:** Due to the closed loop nature of magnetic field lines, each line that enters a surface must also escape at another point,

resulting in a lack of overall flux across the surface (Singh, 2006):

$$\oint_{surface} B \cdot dA = 0$$

Therefore:

$$\oint_{surface} B \cdot dA = 0$$

4.2.2.2. Differential Form of Gauss's Law for Magnetism

Gauss's Law, in its differential form, establishes a relationship between the divergence of the magnetic field and the nonexistence of magnetic monopoles. It is represented as:

$$\nabla \cdot B = 0$$

where, $(\nabla \cdot B)$ represents magnetic field divergence (Field, 2006).

4.2.2.2.1. Derivation of Differential Form

1. Gauss's Law for Magnetism:

$$\oint_{surface} B \cdot dA = 0$$

2. **Apply the Divergence Theorem:** The divergence theorem states that the flux of a vector field through a closed surface is equal to the volumetric integral of the divergence of that field inside the surface:

$$\oint_{surface} B \cdot dA = \int_{volume} (\nabla \cdot B) dV$$

3. Given that the total flow is zero for any given volume, it follows that the integrand must be zero over the whole volume:

$$\int_{volume} (\nabla \cdot B) dV = 0$$

4. Given that the volume integral is 0 for any given volume:

$$\nabla \cdot B = 0$$

Therefore, the differential version of Gauss's Law is obtained, which suggests that the lines of the magnetic field do not have a distinct starting or ending point (Tarasov, 2008).

4.2.3. Faraday's Law of Induction

This law is a fundamental concept in electromagnetism that explains how a magnetic field that changes over time can induce an EMF in a circuit. This principle serves as the foundation for several technical applications, such as transformers, electric generators, and inductors.

It states that the EMF produced inside a closed circuit is exactly equal to the negative of the derivative of the magnetic flux through the circuit. Mathematically, the concept is represented as:

$$E = -\frac{d\Phi_B}{dt}$$

where, (E) is the EMF; and (Φ_B) is the magnetic flux.

The negative sign denotes the orientation of the induced EMF in accordance with Lenz's Law, which asserts that the induced EMF will counteract any change in magnetic flux (Guisasola et al., 2013) (Figure 4.4).

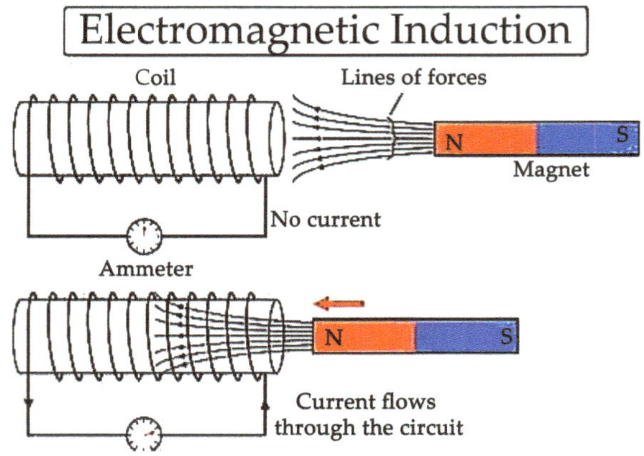

Figure 4.4. *Illustration of the electromagnetic induction.*

Source: Science Facts, Creative Commons License.

4.2.3.1. Integral Form of Faraday's Law

It establishes a connection between the electromotive force created around a closed loop and the rate at which the magnetic flux changes through the area enclosed by the loop. This formula is useful in real-world scenarios, such as determining the electromotive force (EMF) in electrical circuits.

$$\oint_{loop} E \cdot dl = -\frac{d}{dt}\int_{surface} B \cdot dA$$

where, (dl) is the length vector; and (dA) is the differential area vector (Kinsler, 2020).

4.2.3.1.1. Derivation of Integral Form

1. **Imagine a Closed Circuit Made of Wire That is Exposed to a Magnetic Field that Changes Over Time:** As Φ_B across the loop varies over time, an EMF is generated. Φ_B is defined as the total magnetic field traversing through a surface.
2. The Magnetic Flux Passing Through the Surface (A) that is Surrounded by the Loop is:

$$\Phi_B = \int_{surface} B \cdot dA$$

3. Express the Induced EMF:

$$E = -\frac{d\Phi_B}{dt} = -\frac{d}{dt}\int_{surface} B \cdot dA$$

4. Relate EMF to the Electric Field:

$$E = \oint_{loop} E \cdot dl$$

5. Combine the Equations (Zuza et al., 2014):

$$\oint_{loop} E \cdot dl = -\frac{d}{dt}\int_{surface} B \cdot dA$$

4.2.3.2. Differential Form of Faraday's Law

Faraday's Law in its differential version expresses the local correlation between the electric field and the temporal derivative of the magnetic field. It is represented as:

$$\nabla \times E = -\frac{\partial B}{dt}$$

where, $(\nabla \times E)$ is the electric field curl (Barnes & Maekawa, 2007).

4.2.3.2.1. Derivation of Differential Form

1. Start with the Integral Form of Faraday's Law:

$$\oint_{loop} E \cdot dl = -\frac{d}{dt}\int_{surface} B \cdot dA$$

2. **Apply Stokes' Theorem:** It creates a linkage between the line integral of a vector field along a closed loop and the surface integral of the curl of the field over the surface bounded by the loop:

$$\int_{surface} (\nabla \times E) \cdot dA = -\frac{d}{dt}\int_{surface} B\, dA$$

Or,

$$\nabla \times E = -\frac{\partial B}{dt}$$

Therefore, the differential formulation of Faraday's Law is derived, illustrating that a time-varying magnetic field induces a circulating electric field (Zhdanov, 2020).

4.2.4. Ampère's Law (with Maxwell's Addition)

Ampère's Law establishes a relationship between the magnetic field around a closed loop and the electric current flowing through the loop. James Clerk Maxwell subsequently enhanced Ampère's Law by introducing the concept of displacement current, resulting in the version sometimes referred to as Ampère's Law with Maxwell's Addition (Jonson, 2016). Ampère's Law states that the circulation of the magnetic field around a closed loop is directly proportional to the total current flowing through the loop. Mathematically, it is represented using the following notation:

$$\oint_{loop} B \cdot dl = \mu_0 I_{enc}$$

where, (dl) is the differential length vector along the closed loop; (μ_0) is the free space permeability $(\mu_0 \approx 4\pi \times 10^{-7}\ N/A^2)$; and (I_{enc}) is the current enclosed (Jonson, 2015).

4.2.4.1. Maxwell's Addition: Displacement Current

Maxwell's Addition to Ampère's Law adds the idea of displacement current, denoted as (I_d), which considers the changing electric field in instances when there is no actual flow of electric charge. This increase is essential for maintaining the coherence of Maxwell's equations in accurately characterizing electromagnetic waves.

The displacement current is expressed as:

$$I_d = \epsilon_0 \frac{d\Phi_E}{dt}$$

where, (ϵ_0) is the free space permittivity $(\epsilon_0 \approx 8.85 \times 10^{-12}\ F/m)$; and (Φ_E) is the electric flux through the surface (Heras, 2011).

Combining the (I) and (I_d):

$$\oint_{loop} B \cdot dl = \mu_0 \left(I_{enc} + \epsilon_0 \frac{d\Phi_E}{dt} \right)$$

4.2.4.2. Integral Form of Ampère's Law with Maxwell's Addition

The integral version of Ampère's Law with Maxwell's Addition may be obtained by considering the combined effects of both conduction and displacement currents:

$$\oint_{loop} B \cdot dl = \mu_0 \left(I_{enc} + \epsilon_0 \frac{d}{dt} \int_{surface} E \cdot dA \right)$$

where, (dA) is the differential area vector over the surface that is enclosed by the loop.

4.2.4.2.1. Derivation of Integral Form

Consider a closed loop and the surface it encloses:

The surface (S) is bounded by the loop (C).

1. **Define the Conduction Current (I_{enc}):** The current (I_{enc}) passing through the surface (S) is given by (Suárez et al., 2024):

$$I_{enc} = \int_S J \cdot dA$$

where, (J) is the current density.

2. **Include the Displacement Current:** I_d through the surface is:

$$I_d = \epsilon_0 \frac{d}{dt} \int_S E \cdot dA$$

3. **Combine the Contributions:** The conduction current plus displacement current adds up to the total current:

$$\oint_{loop} B \cdot dl = \mu_0 \left(\int_S J \cdot dA + \epsilon_0 \frac{d}{dt} \int_S E \cdot dA \right)$$

Thus, the integral form of Ampère's Law with Maxwell's Addition is derived (Hampshire, 2018).

4.2.4.3. Differential Form of Ampère's Law with Maxwell's Addition

Ampère's Law with Maxwell's Addition, in its differential form, establishes a relationship between the curl of the magnetic field and the combined effect of the current density and the rate of change of the electric field. It is shown as:

$$\nabla \times B = \mu_0 J + \mu_0 \epsilon_0 \frac{\partial E}{dt}$$

where, $(\nabla \times B)$ is the magnetic field curl; and $\left(\frac{\partial E}{dt}\right)$ is the time derivative of the electric field (Tai, 1972).

4.2.4.3.1. Derivation of Differential Form

1. From Ampère's Law with Maxwell's Addition:

$$\oint_{loop} B \cdot dl = \mu_0 \left(I_{enc} + \epsilon_0 \frac{d}{dt} \int_{surface} E \cdot dA \right)$$

2. **Apply Stokes' Theorem:** It establishes a connection between the line integral of a vector field along a closed loop and the surface integral of the curl of the field over the surface enclosed by the loop:

$$\oint_{loop} B \cdot dl = \int_{surface} (\nabla \times B) \cdot dA$$

Combining these equations:

$$\int_{surface} (\nabla \times B) \cdot dA = \mu_0 \left(\int_{surface} J \cdot dA + \epsilon_0 \frac{d}{dt} \int_{surface} E \cdot dA \right)$$

Or,

$$\nabla \times B = \mu_0 J + \mu_0 \epsilon_0 \frac{\partial E}{dt}$$

Hence, the differential expression of Ampère's Law including Maxwell's Addition is obtained (Hill, 2011).

4.3. ELECTROMAGNETIC WAVES

Electromagnetic waves are fundamental phenomena in physics that can travel through a vacuum and other media without requiring a material medium (Bromwich, 1919). These waves are consist of oscillating magnetic and electric fields that are perpendicular to each other and to the direction of wave propagation. This section explores the characteristics of electromagnetic waves and delves into the velocity of light in a vacuum (Stratton & Chu, 1939).

4.3.1. Properties of Electromagnetic Waves

The following are the characteristics of electromagnetic waves.

4.3.1.1. Electric and Magnetic Fields

Electromagnetic waves consist of electric (E) and magnetic (B) fields that are perpendicular to each other. The fields undergo sinusoidal oscillations that are synchronized with each other and are governed by Maxwell's equations (Muhibbullah, 2021).

4.3.1.2. Wavelength and Frequency

The wavelength (λ) of an electromagnetic wave is the spatial distance between consecutive peaks or valleys of the wave. The relationship is inverse between the wave's frequency (f), which denotes the number of oscillations per unit time. Mathematically, the relationship between them may be expressed by the equation ($c = f\lambda$), where (c) represents the speed of light in the medium (Oya, 1971).

4.3.1.3. Propagation Speed

Electromagnetic waves propagate at the speed of light (c) in a vacuum, which is roughly (3×10^8) meters per second (m/s) (Gladyshev et al., 2006). This speed is a fundamental constant in physics. It signifies the highest possible speed at which information or energy may travel across space (Sizun, 2005).

4.3.1.4. Transverse Nature

Contrary to sound waves, which travel longitudinally and require a medium to propagate, electromagnetic waves are transverse. Consequently, the electric and magnetic fields oscillate at right angles to the direction of wave propagation, creating planes that are mutually perpendicular (Rikken & Van Tiggelen, 1996).

4.3.1.5. Energy and Momentum

Electromagnetic waves can transport both energy and velocity. The energy (E) of an electromagnetic wave is directly proportional to its frequency (E=hf), with (h) representing Planck's constant (Dalaklioğlu et al., 2015).

The momentum is given by $\left(p = \frac{E}{c} = \frac{hf}{c} \right)$.

4.3.2. Speed of Light in a Vacuum

The velocity of electromagnetic radiation in a vacuum, often referred to as the speed of light (c), is an immutable and fundamental constant in physics. It is crucial in determining the behavior of electromagnetic waves and forms the foundation of comprehension of space and time (Yener, 2017). The numerical value assigned to the variable (c) is about (3×10^8) meters per second (m/s).

4.3.2.1. Experimental Determination

Ole Rømer achieved the first precise measurement of the speed of light in 1676 by seeing the moons of Jupiter. Subsequent tests conducted by Michelson and Morley during the late 19th century yielded more accurate measurements, verifying that the speed of light remains constant and unaffected by the movement of either the source or the observer.

4.3.2.2. Maxwell's Equations

The equations formulated by James Clerk Maxwell in the field of electromagnetism accurately predicted the existence of electromagnetic waves and established a theoretical framework for understanding their velocity. These equations provide a unified description of the relationship between electricity and magnetism and explain how electric and magnetic fields that are changing in time propagate as waves through space (Bartsch et al., 1992).

4.3.2.3. Implications in Physics

There are enormous implications for the area of physics, particularly for Einstein's theory of relativity, because the speed of light in a vacuum does not change. According to the theory of special relativity, the speed of light is constant for all observers, regardless of the position of the observers in relation to one another. This concept forms the basis for understanding the relationship between space, time, and energy (Wang, 2008).

4.4. ENERGY AND MOMENTUM IN ELECTROMAGNETIC WAVES

Electromagnetic waves transport energy and momentum through space, playing a crucial role in understanding how they interact with matter and their technical uses. This section explores the fundamental principles of energy and momentum in electromagnetic waves, with a specific emphasis on the Poynting vector, radiation pressure, and the intensity of electromagnetic waves (Saldanha, 2011).

4.4.1. Poynting Vector

The Poynting vector (S) is a basic parameter in electromagnetism that characterizes the directional flow of energy (power per unit area) conveyed by an electromagnetic wave. It is characterized by:

$$S = \frac{1}{\mu_0} E \times B$$

where, (μ_0) is the free space permeability $(\mu_0 \approx 4\pi \times 10^{-7}\, N/A^2)$.

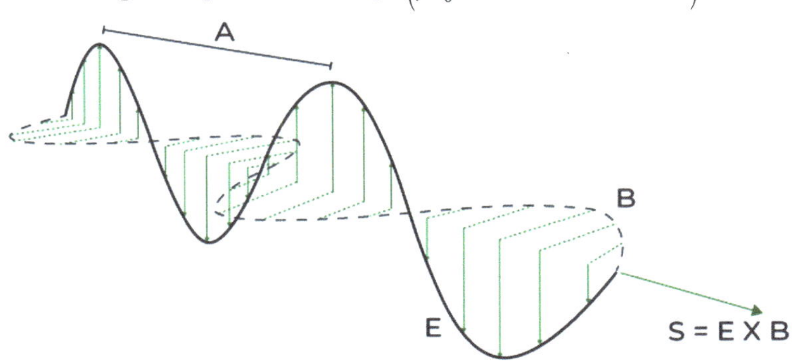

Figure 4.5. *Poynting vector is the cross product of electric field (E) and magnetic field (B).*

Source: Geeks for Geeks, Creative Commons License.

The Poynting vector indicates the direction of energy propagation and displays the rate at which energy flows per unit area per unit time. The magnitude of the electromagnetic wave determines its intensity (Kholmetskii et al., 2016) (Figure 4.5).

4.4.2. Radiation Pressure

Electromagnetic waves apply force to the surfaces they encounter, which is referred to as radiation pressure (P_{rad}). The pressure is caused by the transfer of momentum from photons to the surface. The radiation pressure can be expressed as:

$$P_{rad} = \frac{I}{c}$$

Where, (I) is the intensity of the electromagnetic wave; and (c) is the speed of light.

This equation demonstrates that the radiation pressure is directly proportional to the strength of the electromagnetic wave and inversely proportional to the speed of light (Mansuripur, 2004).

4.4.3. Intensity of Electromagnetic Waves

The intensity (I) of an electromagnetic wave is a measure of the power per unit area carried by the wave and may be mathematically described as:

$$I = \frac{1}{2}\epsilon_0 c E_{max}^2$$

Where, $(\epsilon_0)$is the permittivity of free space $(\epsilon_0 \approx 8.85 \times 10^{-12}\ F/m)$.;$(c)$ is the speed of light in vacuum; and (E_{max}) is the maximum amplitude of the electric field.

Based on the permittivity of free space, this equation shows that the intensity is directly proportional to the square of the maximum amplitude of the electric field. Additionally, the intensity is affected by the properties of the medium, as demonstrated by the characteristics of the medium (Shiryaev, 2008).

4.5. ELECTROMAGNETIC INDUCTION

4.5.1. Inductance

An inherent property of electrical circuits, inductance is governed by Faraday's law of electromagnetic induction. Inductance is a characteristic that cannot be altered. A component, typically a coil or solenoid, is said to have the ability to generate an electromotive force (EMF) when there is a change in the current that is flowing through it or an adjacent coil. This is referred to as the electromotive force characteristic. It is essential that this phenomenon be taken into consideration while designing and operating electrical equipment and systems.

The ratio of the induced electromagnetic field (EMF) to the rate of change of current is the definition of inductance (Thong & Gunstone, 2008):

$$E = -L\frac{di}{dt}$$

Its unit is the Henry (H) which is equivalent to volts per ampere-second $(V / A{\cdot}s)$.

4.5.1.1. Inductors in Circuits

Inductors are passive electrical components specifically designed to provide inductance to a circuit. Inductors are composed of wire coils wrapped around a core material, which may be air, iron, or ferrite. Inductors resist to changes in the flow of electric current, storing energy inside their magnetic fields as current traverses through them. Inductors are used in circuits for various functions, such as energy storage, filtering, and impedance matching.

4.5.1.2. Energy Stored in an Inductor

The stored energy in an inductor is:

$$W = \frac{1}{2}LI^2$$

Where (L) represents the inductance measured in Henrys (H), and (I) denotes the electric current passing through the inductor.

Inductors store energy inside their magnetic fields when electric current passes through them, and then release this energy back into the circuit when the current undergoes alteration (Ferreira & Van Wyk, 2001).

4.5.2. Self-Inductance

Self-inductance refers to the intrinsic property of a coil or solenoid to produce a magnetic field around itself when an electric current passes through it, resulting in inductance. A coil's self-induction is a measure of its capacity to generate an electromotive force inside itself while the electric current passing through it undergoes alteration (Figure 4.6).

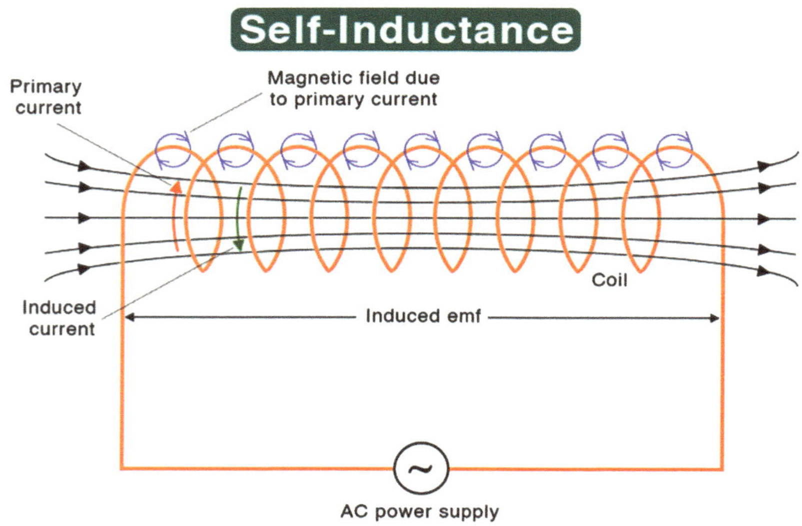

Figure 4.6. *Schematic representation of self-inductance.*

Source: Science Facts, Creative Commons License.

- **Equation and Example:**

The self-inductance of a coil is given by:

$$L = \frac{\Phi_B}{I}$$

Where, (Φ_B) is the magnetic flux.

Example: Calculate the self-inductance of a coil with (100) turns and a magnetic flux of (0.5) Weber when a current of (2) amperes flow through it (Coufal & Radil, 2020).

$$\Phi_B = 0.5Wb$$

$$I = 2A$$

$$L = \frac{0.5Wb}{2A}$$

$$L = 0.25H$$

4.5.3. Mutual Inductance

The electromagnetic phenomenon known as mutual inductance is responsible for explaining the generation of an electromotive force (emf) in one coil because of a change in current in another coil that is positioned near the first coil. Mutual inductance is investigated in this section, which covers topics such as its definition, the units of measurement, the application of mutual inductance in transformers, and the transmission of energy throughout transformers (Figure 4.7).

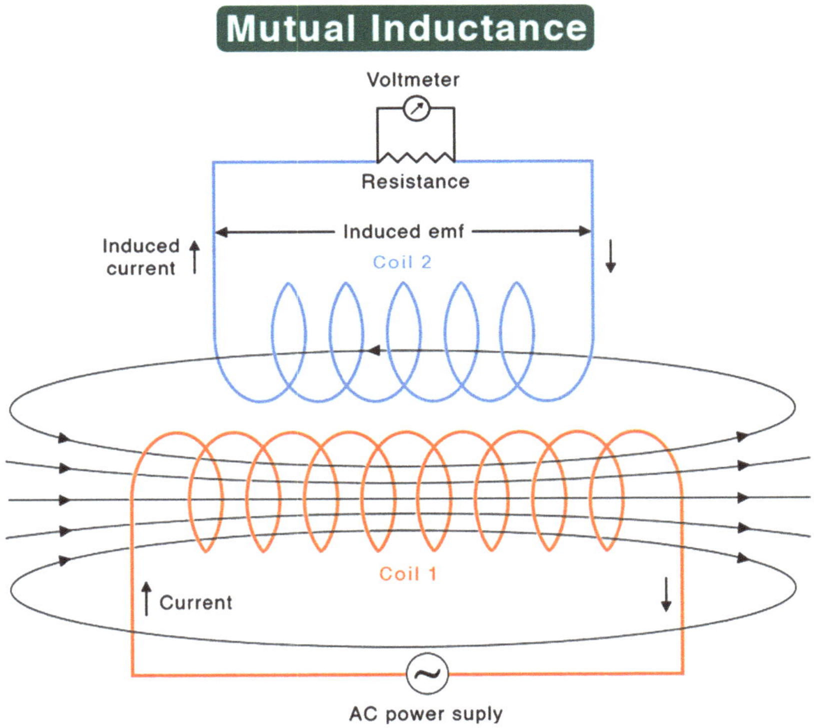

Figure 4.7. *Schematic representation of mutual inductance.*

Source: Science Facts, Creative Commons License.

Mutual inductance is calculated by dividing of the induced electromotive force (emf) in one coil and the rate of change of electric current in another coil positioned near the first one. The expression can be written as follows:

$$M = \frac{\Phi_{B21}}{I_1}$$

Where, (M) is the mutual inductance; and (Φ_{B21}) is the flux passing through the second coil induced by I_1 in the first coil (Rauff et al., 2016).

4.5.3.1. Transformers and Their Applications

Transformers are electrical devices that transfer energy from one circuit to another by utilizing mutual inductance so that the energy can be transferred. Typically, they are consist of two wire coils that are coiled around a magnetic core that is shared by both. In the process of an alternating current (AC) flowing through the main coil, also known as the input coil, a magnetic field that is in a state of constant change is generated. Because of the mutual inductance, the fluctuating magnetic field is responsible for the emf in the secondary coil, which is also known as the output coil.

4.5.3.2. Energy Transfer in Transformers

The concept of mutual inductance is the foundation upon which transformers are built for the transmission of energy. A fluctuating magnetic flux is produced in the core of a transformer whenever alternating current is allowed to flow through the main coil of the transformer. In accordance with Faraday's Law, the changing magnetic field is generates of an emf in the secondary coil. Because of electromagnetic induction, this is the phenomenon that has occurred. The relationship that exists between the transformer and the voltage transformation ratio is established by the ratio of turns that are present between the main and secondary coils of the transformer. In the systems that are responsible for the distribution and transmission of electrical power, transformers are vital components. Voltages are adjusted to suitable levels so that transmission and use can be carried out effectively. Because of the tiny amount of energy that is lost in the magnetic core and windings, transformers have a high energy transfer efficiency that frequently exceeds 95% (Cooper, 1949).

4.5.3.3. Practical Applications

The practical uses are as follows:

1. **Power Transmission:** High-voltage transformers are used to increase voltage levels to facilitate long-distance power transmission, therefore minimizing energy losses and enhancing efficiency.
2. **Voltage regulation:** Transformers are an essential component of power distribution networks, as they are responsible for regulating voltage levels to cater to the respective requirements of homeowners, businesses, and industrial consumers.
3. **Small Transformers** play a crucial role in electronic devices, including power supply, audio equipment, and telecommunications networks.
4. **Instrumentation:** Transformers are used in measuring equipment and sensors to separate and adjust voltage levels for precise measurement and control purposes.

Mutual inductance is a fundamental principle in electromagnetism that underpins the functioning of transformers and their application in electrical power systems and electronics. Understanding mutual inductance is crucial for developing effective and dependable electrical systems for various technological and industrial applications (Evans and Heffernan, 2001).

4.6. ELECTROMAGNETIC WAVES IN COMMUNICATION

Electromagnetic waves are crucial in modern communication technologies since they allow for the efficient transport of information across large distances with minimum waste. This section examines the use of electromagnetic waves in interactions, with a specific emphasis on radio frequencies and antennas, microwave interactions, and fiber optic communication.

4.6.1. Radio Waves and Antennas

Radio waves are a type of electromagnetic radiation that is widely used for wireless networking, television broadcasting, and radar surveillance systems. The wavelengths of these signals span from a few meters to tens of kilometers, which corresponds to frequencies ranging from kilohertz to gigahertz.

- **Equation and Example:**

  The wavelength (λ) of a radio wave is given by:

$$\lambda = \frac{c}{f}$$

Example: A radio station broadcasts at a frequency of $(100)MHz$. Calculate the wavelength of the corresponding radio waves.

$$f = 100 \times 10^{6}\ Hz$$

$$\lambda = \frac{3 \times 10^{8}\ m/s}{100 \times 10^{6}\ Hz}$$

$$\lambda = 3\,meters$$

This calculation shows that the wavelength of (100) MHz radio waves is (3) meters (Norton, 1937).

4.6.2. Microwave Communication

Microwaves are specific type of electromagnetic waves that have wavelengths ranging from one meter to one millimeter, corresponding to frequencies ranging from 300 MHz to 300 GHz. They are often used in communications via satellite, radar systems, and kitchen microwaves because of their capacity to penetrate the atmosphere of the Earth and transmit substantial volumes of data.

Did you know?

Microwave communication utilizes electromagnetic waves with frequencies ranging from 1 GHz to 300 GHz for various applications, including satellite communication and radar.

- **Equation and Example:**

The wavelength of a microwave (λ) is also given by:

$$\lambda = \frac{c}{f}$$

Example: A microwave transmitter operates at a frequency of (5) GHz. Calculate the wavelength of the microwave.

$$f = 5 \times 10^9 \ Hz$$

$$\lambda = \frac{3 \times 10^8 \ m/s}{5 \times 10^9 \ Hz}$$

$$\lambda = 0.06 \, meters$$

This calculation shows that the wavelength of a (5) GHz microwave is (0.06) meters (or (6) centimeters) (Sobol, 1984).

4.6.3. Fiber Optic Communication

Fiber optic communication utilizes optical fibers to transmit data to "via light pulses emitted by lasers or light-emitting diodes (LEDs). These optical fibers consist of either glass or plastic and have the capability to transmit signals over great distances with low signal degradation and large bandwidth.

- **Equation and Example:**

The substance that makes up an optical fiber has a refractive index (n) that determines the speed of light that travels through the fiber:

$$v = \frac{c}{n}$$

Example: Compute the speed of light passing through an optical fiber having a refractive index of 1.5.

$$n=1.5$$

$$v=\frac{3\times10^8\, m/s}{1.5}$$

$$v=2\times10^8\, m/s$$

This calculation shows that light travels at 2×10^8 meters per second in a fiber optic cable with a refractive index of 1.5.

To summarize, electromagnetic waves, such as radio waves, microwaves, and light waves, are essential in modern communication technology. Their characteristics, such as wavelength and frequency, determine their application in various communication systems, ranging from broadcasting and satellite communication to high speed data transfer via fiber optics. Comprehending these characteristics is crucial for creating efficient and dependable communication networks (Sharma et al., 2013).

4.7. ELECTROMAGNETIC INDUCTION IN POWER GENERATION

Electric generators rely on the fundamental concept of electromagnetic induction, which is essential for power production on a global scale. This section explores the use of electromagnetic induction to generate electricity, focusing on electric generators, AC and DC systems, and the distribution of power via the electrical grid.

4.7.1. Electric Generators

In line with the principle of electromagnetic induction, electric generators are devices that can convert mechanical energy into electrical energy. They are consist of a rotor or a coil that is rotating and is placed inside of a magnetic field. The magnetic field causes the coil to generate an electromotive force (emf) or voltage because of the rotation of the coil. A load is connected to the generator, which causes the generator to produce induced voltage, which in turn causes the generator to generate electrical current (Figure 4.8).

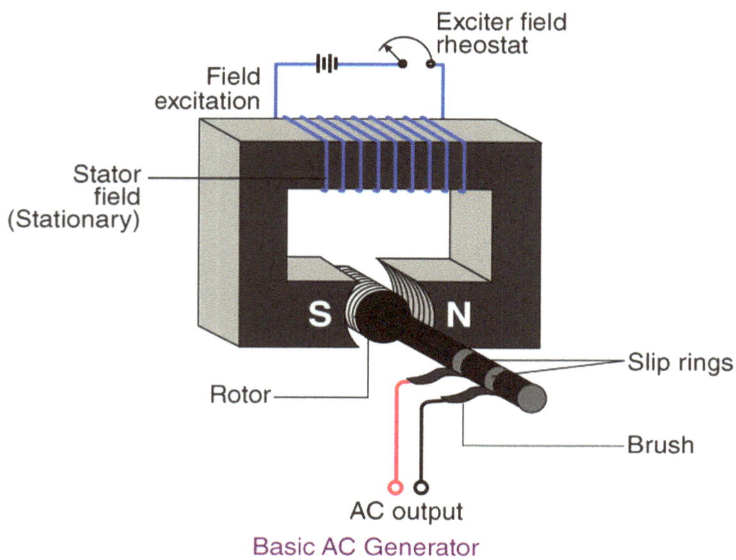

Figure 4.8. *Illustration of a basic AC generator.*

Source: BYJU's, Creative Commons License.

- **Equation and Example:**

The induced emf in a generator coil is given by:

$$E = -N\frac{d\Phi_B}{dt}$$

Where, (E) is the induced emf; and (N) is the turns of the coil.

Example: Calculate the induced emf in a generator coil with $(N=100)$ turns when the magnetic flux changes at a rate of (2) T (tesla) per second.

$$\frac{d\Phi_B}{dt} = 2T/s$$

$$E = -100 \times 2$$

$$E = -200\,volts$$

This calculation shows that the induced emf in the coil is (-200) volts, indicating the direction of the induced current (Boldea, 2017).

Remember

Electric generators are crucial for converting mechanical energy into electrical energy through electromagnetic induction.

4.7.2. Alternating Current (AC) and Direct Current (DC) Systems

Electric generators may generate either AC or DC based on their specific design and intended use. AC generators generate alternating current, which is characterized by the periodic reversal of current direction, usually at a frequency of 50 or 60 Hz. DC generators, in contrast, generate direct current, which exhibits unidirectional movement of electric charge. AC is the favored choice for transmitting electricity over long distances because it can conveniently alter voltage using transformers.

- **Equation and Illustration:**

The frequency of alternating current generated by a generator is given by:

$$f = \frac{n \times P}{120}$$

Where n is the rotational speed and P as the number of poles.

Example: Calculate the frequency of AC produced by a generator running at (1800) RPM with (4) poles.

$$n = 1800\,RPM = \frac{1800}{60} = 30\,Hz$$

$$P=4$$

$$f = \frac{30 \times 4}{120}]$$

$$f = 1\,Hz$$

This calculation shows that the frequency of AC produced by the generator is (1) Hz (Antoniou, 2015).

4.7.3. Grid Distribution

The electrical grid is a network of transmission and distribution wires that transfer electricity produced by power plants. The grid serves as a network that links power generation facilities with end-users, guaranteeing dependable and effective transmission of energy over extensive geographic regions. Transformers are utilized to elevate voltage to facilitate reliable transmission over long distances and to decrease voltage to facilitate efficient distribution to consumers.

- **Equation and Example:**

The power (P) transmitted through a transmission line is given by:

$$P=V \times I$$

Where, (V) is the voltage of the transmission line; and (I) is the current flowing through the transmission line.

Example: Calculate the power transmitted through a transmission line with a voltage of (230) kV and a current of (1000) A.

$$V = 230 \times 10^3\,V$$

$$I = 1000\,A$$

$$P = 230 \times 10^3 \times 1000$$

$$P = 230 \times 10^6\,watts = 230\,MW$$

The calculation indicates that a total power of 230 MW is being delivered across the transmission line.

Electromagnetic induction is the fundamental principle behind the creation of power using electric generators, which produce energy in both AC and DC formats. The electrical grid effectively and systematically transmits electricity from power plants to customers, guaranteeing dependable and extensive access to electrical energy for a broad range of industrial, commercial, and residential uses. It is essential to understand these ideas for the construction and running of contemporary electrical power systems (Medina et al., 2010).

4.8. ELECTROMAGNETIC TECHNOLOGY IN EVERYDAY LIFE

Electromagnetic technology has significantly transformed daily life by facilitating several uses, including induction cooking, wireless charging, and MRI. This part examines the various technologies, delving into their underlying principles and providing concrete instances of how they are used in contemporary culture.

4.8.1. Induction Cooking

Induction cooking uses electromagnetic induction to directly heat cookware, providing fast heating and accurate temperature control. Induction cooktops differ from conventional gas or electric cooktops in that they use electromagnetic fields to produce electrical currents in ferromagnetic cookware, rather than relying on thermal conduction for heat transmission.

For instance: Imagine an induction cooktop equipped with a coil positioned below a glass-ceramic surface. Upon activation, an AC goes through the coil, generating a variable magnetic field. The presence of this field causes the ferromagnetic cookware to produce eddy currents, resulting in the generation of heat via electrical resistance. The cookware rapidly reaches high temperatures, enabling efficient cooking while providing fine control over temperature adjustments (Sweeney et al.,2014).

4.8.2. Wireless Charging

Wireless charging uses electromagnetic induction to transmit electricity without physical connections. This technology enables the transmission of power from a charging pad to various devices, including smartphones, tablets, and wearable gadgets. It avoids the need for physical connections and cords, providing ease and reducing the deterioration of charging ports.

As part of a wireless charging system, a charging pad is comprised of a coil that, when connected to a power source, generates an oscillating magnetic field. A receiver coil is a component of a device that is compatible with the pad. This coil is responsible for collecting the magnetic field and converting it into electrical energy, which is then used to charge the battery of the device. This technology is increasingly being incorporated into various items, including furniture, autos, and public locations, with the goal of providing customers with charging options that are smooth and uninterrupted (Lu et al., 2015).

Tip

Induction cooking is efficient and fast because it uses electromagnetic induction to heat cookware directly, reducing cooking times and energy consumption.

4.8.3. Magnetic Resonance Imaging (MRI)

The magnetic resonance imaging (MRI) technique makes use of intense magnetic fields and radio waves to provide accurate visual representations of the internal structures of the body. This greatly simplifies the process of healthcare diagnosis and the formulation of treatment strategies. It is founded on the principles of nuclear magnetic resonance (NMR), which states that atomic nuclei within the body realign themselves with an external magnetic field and generate radiofrequency signals when they are perturbed by additional magnetic fields (Katti et al., 2011).

When undergoing an MRI scan, the patient is positioned in a reclining position inside a giant cylindrical magnet that generates a strong magnetic field that is stationary. Radiation waves are produced by the radiofrequency coils of the magnetic resonance imaging (MRI) machine. These pulses cause resonance in the hydrogen nuclei (protons) that are present in the tissues of the body. Protons, as they realign themselves, emit signals that are picked up by receiver coils and then transformed into detailed pictures using computer software. These signals are then used to create images. To diagnose diseases like cancer, accidents, and neurological issues, magnetic resonance imaging (MRI) is an indispensable tool.

Ultimately, electromagnetic technology has revolutionized several facets of daily existence, including wireless cooking and device charging, as well as enhancing medical diagnoses via MRI. These applications showcase the adaptability and significance of electromagnetic principles in contemporary technology, improving ease, effectiveness, and healthcare results. Understanding these technologies is crucial for understanding their influence and forthcoming advancements in electromagnetism (Fatahi and Speck, 2015).

ACTIVITY 4.1.

Design and build a simple antenna to explore how radio waves are transmitted and received, demonstrating the principles of electromagnetic wave propagation.

SUMMARY

This chapter explores the fundamental principles and applications of electromagnetic theory. It begins with an overview of the evolution of electromagnetic theory, highlighting key historical developments and foundational concepts that paved the way for modern electromagnetism. The chapter then explores Maxwell's equations, which form the cornerstone of electromagnetic theory. These equations include Gauss's Law for electricity and magnetism, Faraday's Law of Induction, and Ampère's Law with Maxwell's addition. Each law is explained in detail, emphasizing their significance and interrelationships.

Following the discussion of Maxwell's equations, the chapter examines the nature of electromagnetic waves. It describes the properties of these waves and the speed of light in a vacuum. The chapter also addresses the concepts of energy and momentum in electromagnetic waves, focusing on the Poynting vector, radiation pressure, and the intensity of electromagnetic waves. These concepts are crucial for understanding how electromagnetic waves carry and transfer energy. The chapter further explores the principles of electromagnetic induction, covering topics such as inductance, self-inductance, and mutual inductance. These principles are vital for the functioning of various electrical devices and systems. The application of electromagnetic waves in communication technologies is also discussed, including the use of radio waves, microwaves, and fiber optics. These technologies have revolutionized the way we transmit and receive information. In addition, the chapter examines the role of electromagnetic induction in power generation. It explains how electric generators work, the differences between alternating current (AC) and direct current (DC) systems, and the distribution of electricity through the grid. The chapter concludes with a look at the impact of electromagnetic technology on everyday life, discussing applications such as induction cooking, wireless charging, and magnetic resonance imaging (MRI). Through these diverse topics, the chapter provides a comprehensive understanding of electromagnetic theory and its wide-ranging applications.

REVIEW QUESTIONS

1. Describe the historical evolution of electromagnetic theory. What were the key developments that led to the formulation of Maxwell's equations?
2. Explain Gauss's Law for electricity and Gauss's Law for magnetism. How do these laws contribute to our understanding of electric and magnetic fields?
3. What is Faraday's Law of Induction, and how does it explain the process of electromagnetic induction? Provide an example of its application.
7. Discuss the significance of Ampère's Law with Maxwell's addition. How does this law integrate electric currents and changing electric fields in electromagnetism?
8. Define the Poynting vector and radiation pressure. How are these concepts used to describe the energy and momentum in electromagnetic waves?
9. How do inductance, self-inductance, and mutual inductance differ from each other? Provide examples of practical applications for each type of inductance.

10. Explain the role of electromagnetic waves in communication technologies. How do radio waves, microwaves, and fiber optics differ in their applications and properties?

MULTIPLE CHOICE QUESTIONS

1. Which of the following equations is not one of Maxwell's equations?

a. Gauss's Law for Electricity

b. Gauss's Law for Magnetism

c. Coulomb's Law

d. Faraday's Law of Induction

2. The Poynting vector represents which of the following?

a. The electric field strength

b. The magnetic flux density

c. The energy flow per unit area per unit time

d. The potential difference in a circuit

c. What is the primary difference between inductance and self-inductance?

a. Inductance refers to the ability of a coil to induce voltage in another coil, while self-inductance refers to the ability of a coil to induce voltage in itself.

b. Inductance measures resistance, while self-inductance measures capacitance.

c. Inductance only applies to DC circuits, while self-inductance only applies to AC circuits.

d. Inductance is related to electric fields, while self-inductance is related to magnetic fields.

4. In the context of electromagnetic waves, what does the speed of light in a vacuum depend on?

a. The intensity of the electromagnetic wave

b. The frequency of the wave

c. The permittivity and permeability of free space

d. The amplitude of the wave

5. Which communication technology primarily uses fiber optics?

a. Radio broadcasting

b. Microwave transmission

c. Internet data transmission

d. Television broadcasting

Answers to Multiple Questions

1. (c); 2. (c); 3. (a); 4. (c); 5. (c)

REFERENCES

1. Antoniou, D., Tzimas, A., & Rowland, S. M. (2015). Transition from alternating current to direct current low voltage distribution networks. *IET Generation, Transmission & Distribution, 9*(12), 1391–1401.
2. Barnes, S. E., & Maekawa, S. (2007). Generalization of Faraday's law to include nonconservative spin forces. *Physical Review Letters, 98*(24), 246601.
3. Bartsch, M., Dehler, M., Dohlus, M., Ebeling, F., Hahne, P., Klatt, R., & Wolter, H. (1992). Solution of Maxwell's equations. *Computer Physics Communications, 73*(1–3), 22–39.
4. Boldea, I. (2017). Electric generators and motors: An overview. *CES Transactions on Electrical Machines and Systems, 1*(1), 3–14.
5. Brevik, I. (1979). Experiments in phenomenological electrodynamics and the electromagnetic energy-momentum tensor. *Physics Reports, 52*(3), 133–201.
6. Bromwich, T. A. (1919). X. Electromagnetic waves. *The London, Edinburgh, and Dublin Philosophical Magazine and Journal of Science, 38*(223), 143–164.
7. Christodoulou, D. M., & Kazanas, D. (2019). Gauss's law and the source for Poisson's equation in modified gravity with varying G. *Monthly Notices of the Royal Astronomical Society, 484*(1), 1421–1425.
8. Cooper, W. H. B. (1949). An approach to the operation of inductors and transformers in terms of the field impedance. *Proceedings of the IEE-Part II: Power Engineering, 96*(51), 509–517.
9. Coufal, O., & Radil, L. (2020). Self-induction in two long parallel conductors connected to sinusoidal voltage source. *Physica Scripta, 95*(6), 065503.
10. Dalaklioğlu, S., Demirci, N., & Şekercioğlu, A. (2015). Eleventh grade students' difficulties and misconceptions about energy and momentum concepts. *International Journal of New Trends in Education and Their Implications, 6*(1), 13–21.
11. Evans, P. D., & Heffernan, B. (2001). Electromagnetic considerations in power electronic converters. *Proceedings of the IEEE, 89*(6), 864–875.
12. Fatahi, M., & Speck, O. (2015). Magnetic resonance imaging (MRI): A review of genetic damage investigations. *Mutation Research/Reviews in Mutation Research, 764,* 51–63.
13. Ferreira, J. A., & Van Wyk, J. D. (2001). Electromagnetic energy propagation in power electronic converters: Toward future electromagnetic integration. *Proceedings of the IEEE, 89*(6), 876–889.
14. Field, J. H. (2006). Derivation of the Lorentz force law, the magnetic field concept and the Faraday–Lenz and magnetic Gauss laws using an invariant formulation of the Lorentz transformation. *Physica Scripta, 73*(6), 639.
15. Gladyshev, V., Gladysheva, T., & Zubarev, V. (2006). Propagation of electromagnetic waves in media undergoing complex motions. *Journal of Engineering Mathematics, 55*(1), 239–254.

16. Guisasola, J., Zuza, K., & Almudi, J. M. (2013). An analysis of how electromagnetic induction and Faraday's law are presented in general physics textbooks, focusing on learning difficulties. *European Journal of Physics, 34*(4), 1015.
17. Hajnal, J. V. (1987). Singularities in the transverse fields of electromagnetic waves. II. Observations on the electric field. *Proceedings of the Royal Society of London. A. Mathematical and Physical Sciences, 414*(1847), 447–468.
18. Hampshire, D. P. (2018). A derivation of Maxwell's equations using the Heaviside notation. *Philosophical Transactions of the Royal Society A: Mathematical, Physical and Engineering Sciences, 376*(2134), 20170447.
19. Hartree, D. R. (1931). The propagation of electromagnetic waves in a refracting medium in a magnetic field. *In Mathematical Proceedings of the Cambridge Philosophical Society* (Vol. 27, pp. 143–162). Cambridge University Press.
20. Hayakawa, M., Hattori, K., & Ando, Y. (2004). Natural electromagnetic phenomena and electromagnetic theory: A review. *IEEJ Transactions on Fundamentals and Materials, 124*(1), 72–79.
21. Heras, J. A. (2011). A formal interpretation of the displacement current and the instantaneous formulation of Maxwell's equations. *American Journal of Physics, 79*(4), 409–416.
22. Hill, S. E. (2011). Reanalyzing the ampère-maxwell law. *The Physics Teacher, 49*(6), 343–345.
23. Jonson, J. O. (2015). The repulsive force within Ampère's bridge explained by Coulomb's law and special relativity theory, taking into account the effects of propagation delay. *Journal of Basic and Applied Physics, 4*(3), 29–39.
24. Jonson, J. O. (2016). Extended use of Coulomb's law in relation to established laws within electromagnetism. *Journal of Basic and Applied Physics, 5*(1), 40–46.
25. Kaplan, D. B., & Stryker, J. R. (2020). Gauss's law, duality, and the Hamiltonian formulation of U(1) lattice gauge theory. *Physical Review D, 102*(9), 094515.
26. Katti, G., Ara, S. A., & Shireen, A. (2011). Magnetic resonance imaging (MRI)–A review. *International Journal of Dental Clinics, 3*(1), 65–70.
27. Kholmetskii, A., Missevitch, O., & Yarman, T. (2016). Poynting theorem, relativistic transformation of total energy–momentum and electromagnetic energy–momentum tensor. *Foundations of Physics, 46,* 236–261.
28. Kiehn, R. M. (1999). Topological evolution of classical electromagnetic fields and the photon. In V. Dvoeglazov (Ed.), *Photon and Poincaré Group* (pp. 246–262). Nova Science Publishers, Inc.
29. Kinsler, P. (2020). Faraday's law and magnetic induction: Cause and effect, experiment and theory. *Physics, 2*(2), 150–163.
30. Körpınar, T., Demirkol, R. C., Asil, V., & Körpinar, Z. (2022). Magnetic flux surfaces by the fractional Heisenberg antiferromagnetic flow of magnetic b–Lines in binormal direction in Minkowski space. *Journal of Magnetism and Magnetic Materials, 549,* 168952.

31. Lu, X., Wang, P., Niyato, D., Kim, D. I., & Han, Z. (2015). Wireless charging technologies: Fundamentals, standards, and network applications. *IEEE Communications Surveys & Tutorials, 18*(2), 1413–1452.

32. Mansuripur, M. (2004). Radiation pressure and the linear momentum of the electromagnetic field. *Optics Express, 12*(22), 5375–5401.

33. Maries, A., Lin, S. Y., & Singh, C. (2017). Challenges in designing appropriate scaffolding to improve students' representational consistency: The case of a Gauss's law problem. *Physical Review Physics Education Research, 13*(2), 020103.

34. Medina, J., Muller, N., & Roytelman, I. (2010). Demand response and distribution grid operations: Opportunities and challenges. *IEEE Transactions on Smart Grid, 1*(2), 193–198.

35. Muhibbullah, M. (2021). Phase difference between electric and magnetic fields of the electromagnetic waves. *Optik, 247,* 167862.

36. Norton, K. A. (1937). The physical reality of space and surface waves in the radiation field of radio antennas. *Proceedings of the Institute of Radio Engineers, 25*(9), 1192–1202.

37. Oya, H. (1971). Conversion of electrostatic plasma waves into electromagnetic waves: Numerical calculation of the dispersion relation for all wavelengths. *Radio Science, 6*(12), 1131–1141.

38. Poole, G. (2018). Gauss's law for magnetism & law of universal magnetism: Calculate the charge of a monopole. *Journal of High Energy Physics, Gravitation and Cosmology, 4*(4), 581–587.

39. Rauff, K., Rilwan, A., Farouk, U., & Joshua, D. (2016). Construction of a simple transformer to illustrate Faraday's law of electromagnetic induction along side mutual inductance. *Physical Science International Journal, 12*(1), 1–5.

40. Rayleigh, L. (1881). X. On the electromagnetic theory of light. *The London, Edinburgh, and Dublin Philosophical Magazine and Journal of Science, 12*(73), 81–101.

41. Reinhardt, H. (1997). Resolution of Gauss' law in Yang-Mills theory by gauge-invariant projection: Topology and magnetic monopoles. *Nuclear Physics B, 503*(1–2), 505–529.

42. Rikken, G. L. J. A., & Van Tiggelen, B. A. (1996). Observation of magnetically induced transverse diffusion of light. *Nature, 381*(6577), 54–55.

43. Saldanha, P. L. (2011). Division of the energy and of the momentum of electromagnetic waves in linear media into electromagnetic and material parts. *Optics Communications, 284*(12), 2653–2657.

44. Sanders, K., Dappiaggi, C., & Hack, T. P. (2014). Electromagnetism, local covariance, the Aharonov–Bohm effect and Gauss' law. *Communications in Mathematical Physics, 328,* 625–667.

45. Shaarawi, A. M. (2000). Clifford algebra formulation of an electromagnetic charge-current wave theory. *Foundations of Physics, 30*(11), 1911–1941.

46. Sharma, P., Pardeshi, S., Arora, R. K., & Singh, M. (2013). A review of the development in the field of fiber optic communication systems. *International Journal of Emerging Technology and Advanced Engineering, 3*(5), 113–119.

47. Shiryaev, O. B. (2008). Regimes of the interactions of high-intensity plane electromagnetic waves with electron-ion plasmas. *Physics of Plasmas, 15*(1), 87–90.
48. Singh, C. (2006). Student understanding of symmetry and Gauss's law of electricity. *American Journal of Physics, 74*(10), 923–936.
49. Sizun, H. (2005). Electromagnetic waves and propagation characteristics. In *Radio wave propagation for telecommunication applications* (pp. 35–67). Springer.
50. Sobol, H. (1984). Microwave communications-an historical perspective. *IEEE Transactions on Microwave Theory and Techniques, 32*(9), 1170–1181.
51. Stratton, J. A., & Chu, L. J. (1939). Diffraction theory of electromagnetic waves. *Physical Review, 56*(1), 99.
52. Suárez, Á., Marti, A. C., Zuza, K., & Guisasola, J. (2024). Learning difficulties among students when applying Ampére-Maxwell's law and its implications for teaching. *Physical Review Physics Education Research, 20*(1), 10143.
53. Sweeney, M., Dols, J., Fortenbery, B., & Sharp, F. (2014). Induction cooking technology design and assessment. *Small, 5*(2014), 800.
54. Tai, C. T. (1972). On the presentation of Maxwell's theory. *Proceedings of the IEEE, 60*(8), 936–945.
55. Tarasov, V. E. (2008). Fractional equations of Curie–von Schweidler and Gauss laws. *Journal of Physics: Condensed Matter, 20*(14), 145212.
56. Thong, W. M., & Gunstone, R. (2008). Some student conceptions of electromagnetic induction. *Research in Science Education, 38,* 31–44.
57. Wang, X. S. (2008). Derivation of Maxwell's equations based on a continuum mechanical model of vacuum and a singularity model of electric charges. *Progress in Physics, 2,* 111–120.
58. Warnick, K. F., Arnold, D. V., & Selfridge, R. H. (1996). Differential forms in electromagnetic field theory. In *IEEE Antennas and Propagation Society International Symposium. 1996 Digest* (Vol. 2, pp. 1474–1477). IEEE.
59. Yener, N. (2017). Speed of light in vacuum revisited. In *2017 Progress in Electromagnetics Research Symposium-Spring (PIERS)* (Vol. 1, pp. 2378–2383). IEEE.
60. Zhdanov, M. S. (2020). Maxwell's equations and numerical electromagnetic modeling in the context of the theory of differential forms. In *Active Geophysical Monitoring* (Vol. 2, pp. 245–267). Elsevier.
61. Zuza, K., Almudí, J. M., Leniz, A., & Guisasola, J. (2014). Addressing students' difficulties with Faraday's law: A guided problem solving approach. *Physical Review Special Topics-Physics Education Research, 10*(1), 010122.

CHAPTER

5 Solid State Physics

LEARNING OBJECTIVES

After studying this chapter, you will be able to:

- Understand the various types of atomic bonding in solids, including van der Waals attraction and repulsion.
- Explain the structure and properties of various types of crystals, such as ionic crystals.
- Discuss the role of quantum mechanics in atomic bonding within solids.
- Describe the characteristics of crystal structures and the importance of Bravais lattices.
- Apply the concept of Miller indices to identify and describe crystal planes.
- Explore the electronic properties of solids and the application of quantum mechanics to these properties.
- Analyze the phenomena of magnetism and superconductivity, including the Meissner effect and the BCS theory.

KEY TERMS FROM THIS CHAPTER

Band theory of solids	Bravais lattices
Ionic crystals	Lattice dynamics
Meissner effect	Miller indices
Quantum mechanics	Van der Waals attraction

UNIT INTRODUCTION

Solid-state physics is a subfield of condensed matter physics that studies the fundamental characteristics of solids, with a focus on their atomic and electronic structures, as well as their behaviors and interactions under different conditions. The study of this topic involves a thorough exploration of the arrangement of atoms and molecules in crystalline formations, which in turn affects the material's properties, including magnetism, conductivity, and optical characteristics. Solid state physicists aim to clarify the underlying principles that regulate the behaviors of materials by researching phenomena such as phonons (quantized lattice vibrations), band theory, as well as electronic band structures (Ibach & Lüth, 2013).

The practical implications of solid-state physics are extensive and revolutionary in several sectors. Many current advancements rely on the principles of solid-state physics, which include semiconductor devices that power electronics such as computers and smartphones, as well as sophisticated materials utilized in technologies for renewable energy and medical imaging gadgets. Furthermore, ongoing research in this domain is actively pushing the limits, with ongoing investigations into high-temperature superconductivity, innovative magnetic materials for data storage, and quantum materials that hold the potential for breakthroughs in computing and communication technology. Solid-state physics is a fundamental aspect of contemporary science, playing a crucial role in promoting innovation and influencing the technological advancements of the 21st century (Madelung, 2012).

5.1. BINDING

The force that binds atoms together and prevents their collapse is known as the atomic bonding force. The most basic molecule H_2 will be used to explore about the force that keeps it together (Hoffmann, 1987).

5.1.1. Van der Waals Attraction

The cohesive force that holds atoms together and prevents their collapse is referred to as the van der Waals force (Yang et al., 2017) (Figure 5.1).

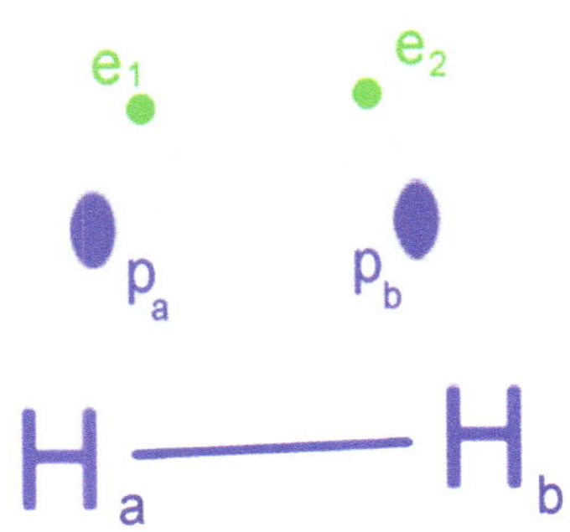

Figure 5.1. *Graphical representation of a* H_2 *molecule.*

Source: Michael Hilke, Creative Commons License.

The entire Hamiltonian may be expressed as the combination of the non-interacting H atoms and the cross terms resulting from Coulomb interactions. Therefore,

$$H = (H_0) + (H_{int}) \tag{1}$$

with

$$H = \frac{\mu}{2m}\left(\vec{r}_1^{\,2} + \vec{r}_2^{\,2}\right) + \frac{e^2}{|\vec{r}_1 - \vec{a}|} + \frac{e^2}{|\vec{r}_2 - \vec{b}|} + \frac{e^2}{\mu}\left(\frac{1}{|\vec{r}_1 - \vec{a}|} + \frac{1}{|\vec{r}_2 - \vec{b}|} - \frac{1}{|\vec{r}_1 - \vec{r}_2|} - \frac{1}{|\vec{r}_1 - \vec{b}|} - \frac{1}{|\vec{r}_2 - \vec{a}|}\right) \tag{2}$$

By using conventional perturbation theory, it becomes feasible to calculate the increase in energy resulting from the presence of $H_{int.}$ The outcome is:

$$\Delta E = -\frac{a \cdot b}{r_{ab}^6} \tag{3}$$

where α_x are the atomic polarizabilities (McLachlan, 1964).

5.1.2. Repulsion

The existence of an attractive potential in the form $-1/r^6$ was recently observed. If the only force acting between the particles in the molecule were a Coulomb repulsion with a strength inversely proportional to the distance between them, the molecule would collapse. Indeed, there exists a very pronounced repulsion that arises from the Pauli exclusion principle. For general purposes, this repulsive potential is often taken to be ~ $1/r^{12}$, which leads to a total potential of the form (Harder & Hu, 2018):

$$\Phi(r) = \frac{-4\sigma}{r^6} + \frac{\sigma}{r^{12}} \quad (4)$$

This is usually referred to as the Lennard-Jones potential. The choice of the repulsive term is rather subjective, although it accurately portrays the limited distance at which the interaction occurs and serves as a reliable approximation to the whole issue (Heiss, 2000).

5.1.3. Crystals

Assuming that the average interatomic distance inside crystals is around 1A°, it may be concluded that (Guyer, 1970):

$$e^2 = \frac{1}{4\pi\epsilon_0}\frac{1}{r} = 14.4eV \quad (5)$$

$$\hbar^2\left(\frac{1}{1\mathring{A}}\right)^2 = 3.8eV \,(6)$$

for the potential in a 1°A quantum box

These energies are much larger when compared to room temperature (300K, 25meV). Therefore, ionic crystals such as NaCl have exceptional stability, characterized by binding energies on the scale of 1 electron volt (eV) (Jones & March, 1985).

5.1.4. Ionic Crystals

Some solids or crystals are primarily bound together by the electrostatic potential, such as alkali-halides like (NaCl –→ Na+Cl–) (Figure 5.2).

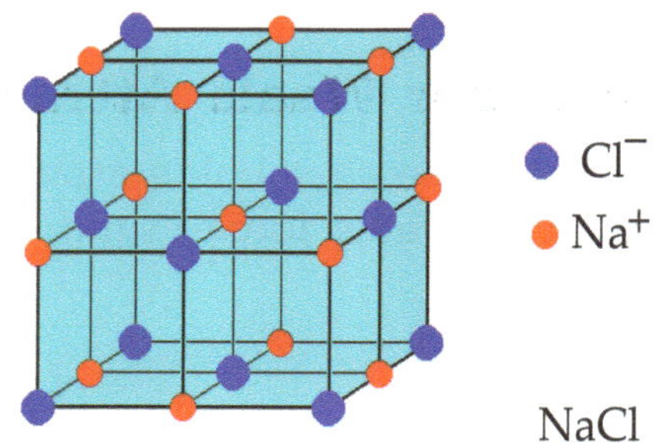

Figure 5.2. *A simple ionic crystal such as NaCl.*

Source: Stack Exchange, Creative Commons License.

The energy per ion pair is

$$\frac{E_{ionpair}}{E} = -\frac{\alpha e^2}{d} + \frac{C}{d^n} = -\frac{\alpha \cdot 14.4eV}{\left[d/\mathring{A}\right]} + \frac{C}{d^n} \quad with\, 6 < n < 12 \,(7)$$

where α is the Madelung constant and can be calculated from the crystal structure. The value of C may be determined by experimental observation of the lowest value in the potential energy. In many cases, the value of n is chosen as 12 to accurately represent the effect of the Pauli exclusion principle. Therefore, by taking the derivative of the potential, one can obtain the following (Tasker, 1979):

$$d_0 = \left(\frac{12C}{e^2\alpha}\right)^{1/11} \quad (8)$$

so that

$$E_{ionpair} = \frac{11\alpha e^2}{12 d_0} \quad (9)$$

5.1.5. Quantum Mechanics as a Bonder

Following is the explanation of quantum mechanics as a bonder:

5.1.5.1. Hydrogen-like Bonding

In a single hydrogen atom, by fixing the proton at position r = 0, one may determine from fundamental principles of quantum mechanics that the energy of the ground state is E^0=-13.6 eV. What is the effect of adding a proton (H^+) to the system that is R distance away? The electron's potential energy is then determined by (Fianchini, 2017):

$$U(r)=\frac{e^2}{r}-\frac{e^2}{|r-R|} \quad (10)$$

The eigenfunction with the lowest eigenvalue of -13.6eV is

$$\xi(r)=\frac{1}{\sqrt{\pi}}\left(\frac{\mu}{a_0}\right)^{3/2} e^{-\frac{r}{a_0}} \quad (11)$$

where a_0 is the Bohr radius. However, there now exists a pair of protons. In the scenario where the protons are infinitely separated, equation (10) may be solved using a linear superposition. This means that the wave function i.e., $\psi(r) = \alpha\xi(r) + \beta\xi(r - R)$, where α and β are coefficients. The lowest eigenvalue of this system is degenerate and has a value of E^0=-13.6 eV. When the value of R is finite, the two eigenfunctions associated with the smallest eigenvalues, E_b and E_a can be approximated by $\psi_b = \xi(r) + \xi(r - R)$ and $\psi_a = \xi(r) - \xi(r - R)$. Refer to Figure 5.3 (Morales et al., 2013).

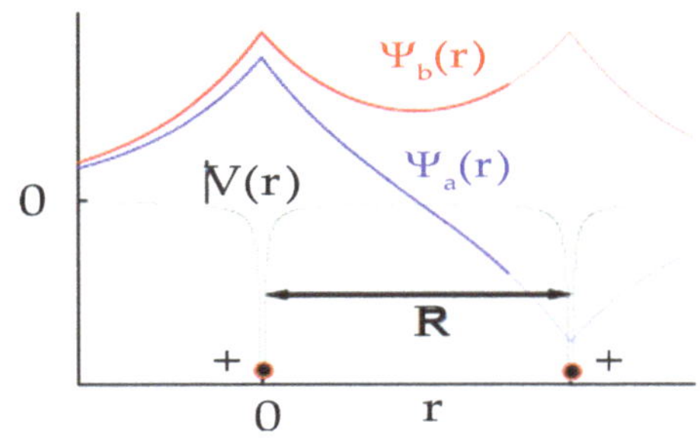

Figure 5.3. *The potential for two protons with the bonding and anti-binding wave function of the electron.*

Source: Michael Hilke, Creative Commons License.

The average energy (or expectation value of E_b) is

$$E_b = \psi_b|H|\psi_b = \psi_b|H|\psi_b \quad (12) \text{ with}$$

$$= E_0 - \frac{A+B}{1+\Delta} \quad (13)$$

$$= e^2 \frac{dr\,\xi^2(r)}{|r-R|} \quad (14)$$

$$B = e^2 \frac{dr\,\xi^2(r)\,\xi}{|r-R|/r} \quad (15)$$

$$\Delta = dr\xi(r)\xi(|r-R|) \quad (16)$$

$$E_a = E^0 - \frac{A-B}{1-\Delta} \quad (17)$$

Hence, the total energy for state ψ_b is now

$$E_b^{total} = E_b + \frac{e^2}{R} \quad (18)$$

When plugging in the numbers E_b^{total} has a minimum at 1.5°*A*.

This diagram demonstrates the rationale for the terms "bonding" and "anti-bonding." In the bonding scenario, the energy decreases as the distance between protons is reduced, provided that $R > R_0$ (Pohl, 1964).

5.1.5.2. Covalent Bonding

Covalent bonds have similarities with hydrogen bonds, but they require the consideration of a linear combination of atomic orbitals for each atom involved. For every N atom, there will be N levels, with the lowest energy state exhibiting a bonding wave function (Raty et al., 2019).

5.1.5.3. Metals

Bonding in metals is a result of the combined actions mentioned above. The concept involves the examination of a group of electrons that are loosely attached to the atomic lattice. The total electrostatic energy may be expressed as (Weiss, 2013):

$$E_{el} = \int dr\, n(r) \sum_{R} \frac{e^2}{|r-R|} + \sum_{R>R} \frac{e^2}{|R-R_0|} + \frac{1}{2} \int dr_1 dr_2 \frac{e^2 n(r_1) n(r_2)}{|r_1 - r_2|} \tag{19}$$

which corresponds to an ionic contribution of the form:

$$E_{el} = \frac{-\alpha e^2}{2r_s}] \quad \text{where,} \quad r_s = \left(\frac{3}{4\pi n} \right)^{1/3} \tag{20}$$

and α is the Madelung constant. Deriving this requires quite a bit of effort. On top of this one must add the kinetic energy of the electrons, which is of the form (Kubota et al., 2018):

$$E_{kin} = \left(\frac{9\pi}{4} \right)^{2/3} \frac{3\hbar^2}{10mr_s^2} \tag{21}$$

Lastly, one must include the exchange of energy, which arises because of the Pauli exclusion principle. The formula for this term is provided by (Herrmann, 1986):

$$E_{ex} = -\left(\frac{9\pi}{4} \right)^{1/3} \frac{3}{4\pi r_s} \tag{22}$$

Putting all this together:

$$E_{ex} = \left(\frac{24.35a_0}{r_s} + \frac{30.1a_0^2}{r_s^2} - \frac{12.5a_0}{r_s} \right) eV / atom \tag{23}$$

The equation results in a minimal value when r_s/a_0 is equal to 1.6. Comparing this with experimental data shows that the outcome deviates by a factor ranging from 2 to 6. What factors caused the discrepancies? This issue was addressed at a semiclassical level, without including all electron-electron interactions in a quantum theory. This task is very challenging, but in scenarios with high density, it is possible to make an estimate and get a more accurate alignment with experimental results (Schwarz and Johnson, 1983).

5.2. CRYSTAL STRUCTURES AND LATTICE DYNAMICS

Crystal structures are the foundation of solid-state physics, determining the material characteristics and behaviors essential for many applications in industry and research. Gaining a comprehensive understanding of crystal structures is crucial in several domains, including materials science and semiconductor device production. Various crystal lattices define the spatial configuration of atoms or ions in a crystal, with each lattice displaying distinct qualities and behaviors stemming from its structural arrangement (Gonze et al., 2005).

5.2.1. Crystal Structures

Crystal structures are characterized by the precise and organized positioning of atoms or ions in a repetitive three-dimensional pattern observed in crystalline materials. The arrangement of atoms in a crystal lattice is controlled by its internal symmetry, which in turn influences the material's chemical and physical properties. The investigation of crystal structures is crucial as it provides a valuable insights of the mechanical, optical, electrical, and thermal properties of a material, which in turn affect its performance in various settings and applications (Müller, 2009).

Crystal structures are characterized by the regular arrangement of component particles, such as atoms, ions, or molecules, inside a solid crystal. The periodicity of the crystal lattice results in anisotropy, which means that the characteristics of the crystal vary in direction. It also results in coherence in the propagation of waves, such as X-rays. The significance of crystal structures lies in their direct association with material characteristics and their impact on technological progress, including the creation of novel materials for electronics, photonics, and renewable energy technologies (Choe et al., 2011).

Crystal lattices are categorized into several types according to their symmetry and atomic structures:

1. **Cubic Lattice:** Defined by angles of 90 degrees and equal lengths along all crystallographic axes. Body-centered cubic (BCC), face-centered cubic (FCC), and simple cubic structures are among them.

2. **Hexagonal Lattice:** It is characterized by a hexagonal unit cell with two axes at 120 degrees and a third axis perpendicular to the plane.
3. **Tetragonal Lattice:** Resembling a cubic lattice, but with varying lengths along one axis.
4. **Orthorhombic Lattice:** It is characterized by three axes of varying lengths and angles of 90 degrees.
5. **Monoclinic Lattice:** It is characterized by two axes that are perpendicular to each other, while the third axis is tilted.
6. **Triclinic Lattice:** It is characterized by three axes of varied lengths and uneven angles (Gavezzotti, 1994).

The physical and chemical properties of a material, such as electrical conductivity, thermal conductivity, and mechanical strength, are influenced by the specific crystal lattice it has. An in-depth understanding of crystal lattices is essential for comprehending how materials behave under various conditions and for customizing material characteristics to suit specific applications (Paulus et al., 2007).

Tip

Understanding different types of Bravais lattices (e.g., simple cubic, face-centered cubic) helps in visualizing crystal structures, aiding comprehension of solid-state physics concepts.

5.2.2. Bravais Lattices

The 14 different three-dimensional configurations that define the periodic structure of crystals are represented by Bravais lattices. Auguste Bravais, a French scientist, gave his name to these lattices, which represent the basic units of crystal formations. In these structures, each lattice point is surrounded by identical environment. Understanding Bravais lattices is crucial for examining the precise placement of atoms in crystalline solids, which in turn affects their physical properties and technological applications (Andrews and Bernstein, 1988).

5.2.2.1. Characteristics and Classification

Bravais lattices are classified according to the symmetry and dimensions of their crystallographic axes. They are categorized into seven crystal systems, with each system consisting of one or more Bravais lattices (Figure 5.4).

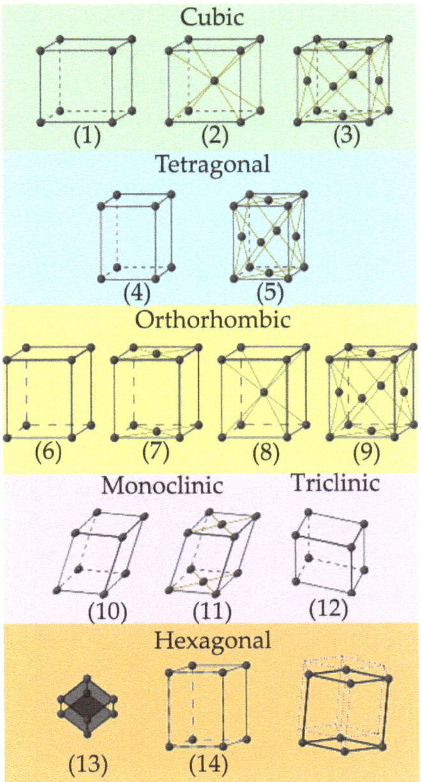

Figure 5.4. *Crystalline materials fit into one of fourteen recognized lattice arrangements. These are known as Bravais lattices.*

Source: Chemicool, Creative Commons License.

The 14 Bravais lattices consist of:

a. Cubic System:
   a. **Simple Cubic (SC):** Atoms are positioned at every vertex of the cube.
   b. **Body-Centered Cubic (BCC):** An extra atom positioned in the middle of the cube.
   c. **Face-Centered Cubic (FCC):** Atoms are located at each vertex and in the middle of each face of the cube.

b. Tetragonal System:
   a. **Simple Tetragonal:** Like a standard SC (single crystal), but with varying lengths along one axis.
   b. **Body-Centered Tetragonal:** The BCC variation has varying lengths along a single axis.

c. Orthorhombic System:
   a. **Simple Orthorhombic:** All three axes have varying lengths and are positioned at 90-degree angles to each other (Ono et al., 2005).
   b. **Base-Centered Orthorhombic:** Extra atoms located in the central positions of two opposing sides.
   c. **Body-Centered Orthorhombic:** One extra atom located in the central position of the unit cell.
   d. **Face-Centered Orthorhombic:** Additional atoms at the centers of all faces.

11. Hexagonal System:
   a. **Hexagonal:** The unit cell consists of six atoms grouped in a hexagonal configuration.

12. Trigonal (Rhombohedral) System:
   a. **Trigonal:** A configuration of three axes that connect at angles that are not 90 degrees.

13. Monoclinic System:
   a. **Simple Monoclinic:** There are two axes that intersect at a 90-degree angle, and there is one inclined axis.
   b. **Base-Centered Monoclinic:** Additional atoms at the center of two opposite faces.

14. Triclinic System:
   a. **Triclinic:** All three axes of different lengths and all

angles are unequal (Moeck & DeStefano, 2018).

5.2.2.2. Unit Cells and Crystal Systems

The unit cell is the most fundamental repeating unit that defines the whole crystal structure. It represents the symmetry and proportions of the lattice structure. The dimensions of the unit cell are defined by lattice parameters, which include the lengths of its edges *(a, b, c)* and the angles formed between them *(α, β, γ)* (Figure 5.5).

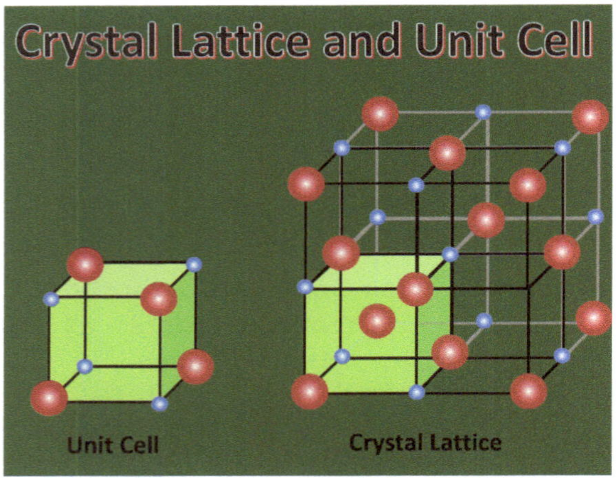

Figure 5.5. *Crystal lattice and unit cell.*

Source: Expii, Creative Commons License.

1. The cubic system is characterized by having all edges of equal length and all angles measuring 90 degrees. Examples include SC, BCC, and FCC.
2. The tetragonal system is characterized by having two edges of identical length and one edge of a different length, with all angles measuring 90 degrees. Examples: Simple Tetragonal, Body-Centered Tetragonal.
3. The orthorhombic system is characterized by having all edges of varying lengths and all angles measuring 90 degrees. Examples of orthorhombic crystal structures include the Simple Orthorhombic, Base-Centered Orthorhombic, Body-Centered Orthorhombic, and Face-Centered Orthorhombic (Donnay & Harker, 1937).
4. The hexagonal system consists of two edges that are equal in length and one edge that is different. The angles between the equal edges measure 120 degrees. Example: Hexagonal.
5. The trigonal (rhombohedral) system consists of edges that are all equal, with angles that are not 90 degrees. Example: Trigonal.
6. The monoclinic system is characterized by having all edges of different lengths, two angles measuring 90 degrees, and one angle that is distinct from the others. Examples: Simple Monoclinic, Base-Centered Monoclinic.
7. The triclinic system is characterized by having all edges of various lengths and all angles unequals. Example: Triclinic crystal.

Comprehending Bravais lattices and unit cells provides a fundamental basis for investigating the arrangement of atoms in solids and their influence on material characteristics. Acquiring this understanding is essential for creating and evaluating materials for a wide range of technical applications, such as electronics, photonics, and structural engineering (Pitteri and Zanzotto, 1996).

5.2.3. Crystal Planes and Miller Indices

Understanding crystal planes is crucial to grasp the geometrical and physical characteristics of crystalline materials. The material's durability, optical behavior, and electrical characteristics are significantly influenced by these planes, which are characterized by the periodic arrangement of atoms. Miller indices provide a systematic approach to identifying and defining certain planes within the crystal lattice, which aids in the study and effective transmission of crystallographic information (Sun et al., 2020) (Figure 5.6).

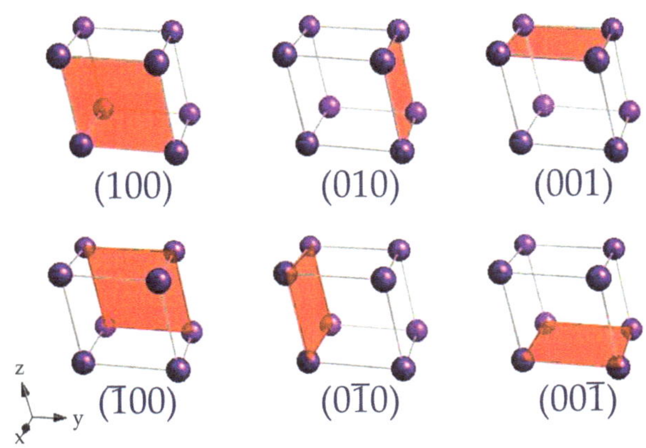

Figure 5.6. *Examples of lattice planes.*

Source: University of Cambridge, Creative Commons License.

Crystal planes are defined as the planar surfaces within a crystal that intersect with lattice points. These planes play a crucial role in understanding the mechanisms of crystal cleavage, growth, and their interactions with external forces and fields. The configuration and alignment of crystal planes significantly impact on the scattering of X-rays, the movement of electrons in semiconductors, and the mechanisms of plastic deformation (Blakely & Somorjai, 1977).

Miller indices are a method of notation used to precisely define the orientation of crystal surfaces and directions inside a crystal lattice. The Miller indices (hkl) are calculated as the reciprocals of the intercepts made by a plane with the crystallographic axes. The Miller indices (hkl) represent the intercepts of a particular plane with the x, y, and z axes, denoted as $(a/h, b/k,)$ and (c/l) accordingly. When a plane is parallel to an axis, the intercept for that axis is considered to be infinitely far away, and the corresponding Miller index is zero (Tian et al., 2008).

5.2.3.1. Determination of Miller Indices

The process of determining Miller indices involves several steps:

1. **Identify the Intercepts:** Calculate the points where the plane crosses the x, y, and z axes. If the plane intersects the origin, either translate the origin or choose a parallel plane that is equal.
2. **Take Reciprocals:** Take the reciprocals of the intercepts along the x, y, and z axes. For instance, if the intercepts are $(a, b,)$ and (c) along the x, y, and z axes, the reciprocals are $\left(\frac{1}{a}, \frac{1}{b}, \frac{1}{c}\right)$.
3. **Clear Fractions:** Multiply the reciprocals by a common factor to obtain the smallest set of integers. These integers are the Miller indices (hkl).

Example:

Consider a plane that intersects the x-axis at (a), the y-axis at $(b/2)$, and is parallel to the z-axis (intercepts at infinity). The steps to determine the Miller indices are:

1. Intercepts: $(a, b/2, \infty)$;
2. Reciprocals: $\left(\frac{1}{a}, \frac{2}{b}, \frac{1}{\infty}\right) \rightarrow \left(\frac{1}{a}, \frac{2}{b}, 0\right)$;
3. Clear Fractions: Multiply by (ab) (the common factor) $\rightarrow ((b, 2a, 0))$.

Thus, the Miller indices are (120) (Sangian et al., 2019).

5.2.3.2. Equation for Miller Indices Determination

Given intercepts at $((x_1, y_1, z_1))$ along the x, y, and z axes, the Miller indices $((h, k, l))$ are determined as:

$$h = \frac{1}{x_1}, \quad k = \frac{1}{y_1}, \quad l = \frac{1}{z_1}$$

Miller indices are of significant importance in several areas, such as the examination of crystal symmetry, the investigation of crystal defects, and the understanding of diffraction patterns in X-ray crystallography. Miller indices are essential tools in the fields of solid-state physics and material studies as they provide a precise and consistent method for describing crystal planes (Aray et al., 2009).

5.2.4. Lattice Dynamics

Lattice dynamics is a fundamental field in solid-state physics that investigates the vibrational characteristics of atoms in a crystal lattice. Understanding these oscillations is crucial for elucidating various physical properties of substances, including thermal conductivity, specific heat, and electrical conductivity. Phonons, which are the quantized modes of vibrations in the crystal lattice, play a significant role in this context. Theoretical models, such as the Einstein and Debye models, offer frameworks for examining these lattice vibrations (Ackland et l., 1997).

5.2.4.1. Phonons and Lattice Vibrations

Phonons are discrete units of vibrations in a lattice structure, like photons in the realm of light. They are coherent oscillations that propagate through the ordered arrangement of atoms in the crystal lattice as wave-like disturbances.

The oscillations of atoms inside a crystal can be described as a combination of normal modes, each characterized by a distinct wave vector (k) and frequency (ω). The relationship between the wave vector and the frequency is known as the phonon dispersion relation.

The vibrational modes can be classified into acoustic and optical phonons:

1. **Acoustic Phonons:** These modes are associated with vibrations in which atoms move in synchrony with one other, leading to the transmission of sound waves. The frequency of acoustic phonons decreases as the wave vector (k) approaches zero.
2. **Optical Phonons:** These modes include vibrations in which atoms in the basis oscillate in a manner where they are out of sync with one other, usually happening at higher frequencies. They play a crucial role in comprehending the interaction between light and the crystal lattice.

The phonon dispersion relation for a one-dimensional monatomic lattice may be obtained by analyzing the equation of motion governing the atoms:

$$M\ddot{u}_n = K\left(u_{n+1} + u_{n-1} - 2u_n\right)$$

where (M) is the mass of an atom, (K) is the spring constant, and (u_n) is the displacement of the nth atom (Harper et al., 1973).

5.2.4.2. Einstein and Debye Models

The Einstein and Debye models are two theoretical approaches used to describe the heat capacity of solids at different temperatures.

5.2.4.2.1. Einstein Model

The Einstein model assumes that each atom in a crystal vibrates independently with the same frequency ( ) The heat capacity is derived by considering the quantization of vibrational energy levels:

$$lE_n = \left(n + \frac{1}{2} \right) \hbar\omega_E$$

The Einstein heat capacity (C_E) is given by:

$$C_E = 3Nk_B \left(\frac{\theta_E}{T} \right)^2 \frac{e^{\theta_E/T}}{\left(e^{\theta_E/T} - 1\right)^2}$$

where (N) is the number of atoms, (k_B) is the Boltzmann constant, $(\theta_E = \hbar\omega_E / k_B)$ is the Einstein temperature, and (T) is the absolute temperature (Fagnocchi et al., 2010).

5.2.4.2.2. Debye Model

The Debye model considers the vibrational modes of the entire crystal lattice, treating the vibrations as a continuum up to a maximum frequency (Debye frequency (ωD). The heat capacity in the Debye model is given by:

$$C_D = 9Nk_B \left(\frac{T}{\theta_D} \right)^3 \int_0^{\theta_D/T} \frac{x^4 e^x}{\left(e^x - 1\right)^2} dx$$

where (θ_D) is the Debye temperature, representing the maximum vibrational frequency.

Both models provide valuable perspectives on the relationship between temperature and heat capacity in solids. The Einstein model is more precise at high temperatures, but the Debye model provides a more exact description of the behavior at low temperatures, where the heat capacity changes according to the cube of the temperature (T^3).

A comprehensive understanding of lattice dynamics and its corresponding phonon characteristics is crucial for elucidating diverse thermal and electrical phenomena in solid materials, hence influencing the formulation and efficacy of materials in technology and industry (Lazebnik et al., 2007).

5.3. ELECTRONIC PROPERTIES AND BAND THEORY

Understanding the electrical, thermal, and optical characteristics of materials requires a basic knowledge of their electronic properties. Band theory, a fundamental principle in the field of solid-state physics, elucidates how electrons fill energy bands that arise from the merging of atomic orbitals within a crystal lattice. The electronic band structure of a material determines whether it is an insulator, conductor, or semiconductor. Conductors have valence and conduction bands that overlap, facilitating unrestricted electron mobility. Semiconductors possess a narrow band gap that allows electrons to traverse under certain circumstances, while insulators possess a wide band gap that obstructs the movement of electrons. Modern electronic gadgets and technologies, such as transistors, diodes, and solar cells, are based on this principle (Herman, 1958).

5.3.1. Electronic Properties

The study of the electronic characteristics of materials is essential for understanding their behavior and their applications in many technological fields. The categorization of substances into conductors and insulators, as well as semiconductors, is determined by their capacity to facilitate the flow of electric current. This categorization is based on band theory, which elucidates the distribution of electrons in energy bands inside a crystal lattice (Sólyom, 2008).

5.3.1.1. Conductors

Conductors are substances that facilitate the unrestricted movement of electric current. This phenomenon occurs due to the overlapping of the valence band and conduction band, or the presence of a minimal energy gap between them, which enables the free movement of electrons. Copper, aluminum, and silver are common examples of metals that can conduct electricity. The electrical conductivity (σ) of a conductor can be expressed by the Drude model:

$$\sigma = n e \mu$$

where (n) is the number of charge carriers per unit volume, (e) is the elementary charge, and (μ) is the mobility of the charge carriers (Patsalas et al., 2018).

5.3.1.2. Insulators

Insulators are substances that, under normal circumstances, do not allow electric current to flow through them due to a significant difference in energy levels between the valence band and the conduction band. The band gap, generally exceeding 4 eV, inhibits the ability of electrons in the valence band to acquire sufficient energy to transition to the conduction band. Typical insulators include rubber, glass, and ceramics. The electrical resistivity (ρ) of an insulator is defined as (Abanin and Pesin, 2011):

$$\rho = \frac{1}{\sigma}$$

5.3.1.3. Semiconductors

Semiconductors possess an energy band gap that falls between the band gaps of insulators and conductors, generally ranging from 0.1 to 4 eV. Under certain circumstances, such as the introduction of impurities or the application of external energy like heat or light, these materials can carry an electric current. Both germanium and silicon are prototypical examples of semiconductors. The conductive ability of semiconductors is contingent upon temperature and may be characterized by:

$$\sigma(T) = \sigma_0 e^{-\frac{E_g}{2k_BT}}$$

where (σ_0) is a material-specific constant, (E_g) is the band gap energy, (k_B) is the Boltzmann constant, and (T) is the temperature (Panich, 2008).

5.3.1.4. Intrinsic Versus Extrinsic Semiconductors

5.3.1.4.1. Intrinsic Semiconductors

Intrinsic semiconductors are pure materials without any significant impurities. Their electrical conductivity is solely due to the electrons that gain enough thermal energy to jump from the valence band to the conduction band, leaving behind holes in the valence band. The number of electrons (n_i) and holes (p_i) in an intrinsic semiconductor is equal and is given by:

$$n_i = p_i = \sqrt{N_c N_v} e^{-\frac{E_g}{2k_BT}}$$

where (N_c) and (N_v) are the effective density of states in the conduction and valence bands, respectively (Stuke, 1970).

5.3.1.4.2. Extrinsic Semiconductors

Impurities are added to extrinsic semiconductors in order to enhance their electrical conductivity. These semiconductors may be classed as either n-type or p-type, depending on the kind of impurity that is introduced.

- **n-type Semiconductors:** The material is infused with donor atoms, which contribute more electrons, so augmenting the population of conduction electrons. Electrons are the prevailing carriers.

$$[n \approx N_D]$$

where (N_D) is the donor concentration.

- **p-type Semiconductors:** Doped with acceptor atoms that create holes in the valence band, increasing the number of holes. The majority of carriers are holes.

$$[p \approx N_A]$$

where (N_A) is the acceptor concentration.

Comprehending these fundamental principles is crucial for the development and enhancement of electronic devices, ranging from basic diodes to intricate integrated circuits. Semiconductor technology and physics are based on the study of how charge carriers behave in various materials under different conditions (Odobel and Pellegrin, 2013).

5.3.2. Quantum Mechanics in Solids

Quantum mechanics is essential for understanding the behavior of electrons in materials. Wave-particle duality, along with the equation developed by Schrödinger are fundamental for studying the electrical characteristics of crystalline materials.

5.3.2.1. Wave-Particle Duality

The idea of wave-particle duality refers to the phenomenon in which particles, like electrons, exhibit characteristics of both waves and particles. The concept of duality is essential in quantum physics and is explained by the de Broglie hypothesis, which asserts that every moving particle or object is associated with a corresponding wave. The de Broglie wavelength (λ) of a particle is given by:

$$e\lambda = \frac{h}{p}$$

where (h) is Planck's constant and (p) is the momentum of the particle. For an electron with momentum (p), this relationship shows that the electron has wave-like properties with a wavelength that can be observed in diffraction experiments.

The wave-like behavior of electrons in solids is crucial as it gives rise to the formation of energy bands and the band structure of materials. The Schrödinger equation is used to elucidate and anticipate the energy states of electrons in a crystal lattice, therefore characterizing their wave-like behavior (Chang and Lee, 2015).

5.3.2.2. Schrödinger Equation in Crystals

The Schrödinger equation is a fundamental equation in quantum mechanics that governs the temporal evolution of the quantum state of a physical system. When studying solids, scientists often use the time-independent Schrödinger equation to examine how electrons behave inside a crystal lattice.

$$-\frac{\hbar^2}{2m}\nabla^2 \psi(r) + V(r)\,\psi(r) = E\,\psi(r)$$

Where, $(\hbar)$ is the reduced Planck's constant; (m) is the electron mass; (∇^2) is the Laplacian operator; $(\psi(r))$ is the wave function of the electron; $(V(r))$ is the potential energy; and

(E) is the energy eigenvalue.

In a crystal lattice, the potential energy $(V(r))$ is periodic due to the regular arrangement of atoms. This periodic potential can be expressed as:

$$V(r+R)=V(r)$$

where (R) is a lattice vector. The solutions to the Schrödinger equation in a periodic potential are the Bloch functions, which can be written as (Christianto, 2014):

$$\psi_k(r)=e^{\pm \cdot} u_k(r)$$

where (k) is the wave vector, and $(u_k(r))$ is a function with the same periodicity as the crystal lattice. The Bloch theorem states that the wave functions in a periodic potential can be expressed as a plane wave modulated by a periodic function.

The energy eigenvalues $(E(k))$ form continuous energy bands due to the periodic potential. These bands elucidate the electrical characteristics of materials, including their conductivity as well as band gaps. The concept of band theory emerges from the solution of the Schrödinger equation in a periodic potential, resulting in the categorization of materials as conductors, insulators, or semiconductors according to their band structure.

Ultimately, understanding the behavior of electrons in materials requires a firm grasp of wave-particle duality and the Schrödinger equation. The presence of a periodic potential inside a crystal lattice results in the creation of energy bands, which dictate the electrical characteristics of the material. These ideas serve as the basis for the development of modern electrical devices and technology (Zhong & Zhao, 2013).

5.3.3. Band Theory of Solids

The Band Theory of Solids is essential for understanding the electrical characteristics of materials. This text elucidates the process by which the discrete energy levels of atoms merge into energy bands within a solid, hence determining the material's conductive, insulating, or semiconducting properties (Chamel, 2012).

Did you know?

Band theory explains how electrons behave in a solid, forming energy bands and bandgaps crucial for understanding electrical conductivity and other electronic properties.

5.3.3.1. Energy Bands and Band Gaps

Electrons in individual atoms are found at distinct energy levels. When atoms combine to create a solid, the electron orbitals in their outer shells overlap because of their close proximity. The overlapping of these discrete energy levels results in the splitting of the levels into several closely spaced levels, which then form continuous energy bands. The crucial bands to consider are the valence band and the band of conduction (Figure 5.7).

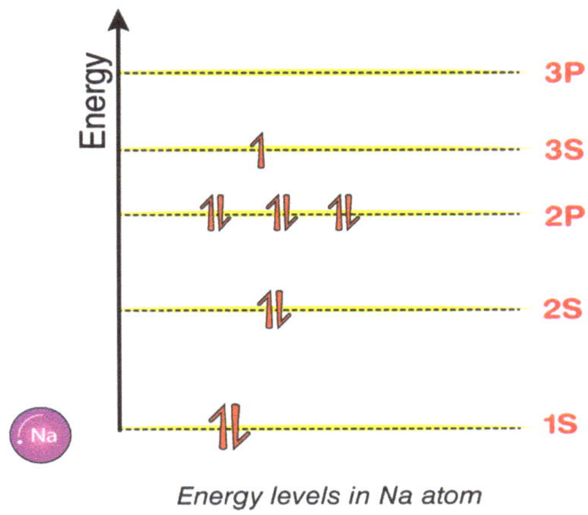

Figure 5.7. *Energy band inside an atom.*

Source: BYJU's, Creative Commons License.

1. **Valence Band:** The energy band formed by the outermost electrons that are involved in chemical bonding.
2. **Conduction Band:** The energy band formed by the electrons that are free to move and contribute to electrical conduction.

The energy gap between the valence band and the conduction band is known as the band gap. The size of the band gap determines the material's electrical properties (Perdew et al., 2017).

5.3.3.2. Origin of Band Formation

The formation of energy bands in solids can be understood by analyzing the overlap of atomic orbitals when atoms combine to form a solid. The description of this process may be elucidated by using the principles of molecular orbitals and crystal field theory.

1. **Atomic Orbitals Overlap:** When atoms are brought into proximity to one another to create a solid, their atomic orbitals intersect. The overlapping of atoms results in the splitting of discrete energy levels into a number equivalent to the number of atoms present. With an increasing number of atoms, there is a corresponding rise in the amount of energy levels, resulting in the formation of continious energy bands.
2. **Pauli Exclusion Principle:** The Pauli Exclusion Principle states that it is impossible

for two electrons inside an atom to possess identical quantum numbers. The exclusion principle is maintained by causing the energy levels to divide and form bands, ensuring that all electrons may be accommodated without any violation.

3. **Energy Level Splitting:** The energy level splitting may be comprehended by using the tight-binding concept. When two atomic orbitals intersect, they create bonding and anti-bonding molecular orbitals with distinct energy levels. Within a solid containing a significant quantity of atoms, this division leads to the creation of energy bands.
4. **Kronig-Penney Model:** This model provides a quantum mechanical foundation for the formation of energy bands. This model examines the behavior of electrons in a one-dimensional periodic potential. It demonstrates that the energy levels that electrons may occupy form bands, with gaps where no energy levels are permitted.
5. **Density of States:** The density of states (DOS) function describes the number of electronic states at each energy level that is available to be occupied. In a solid, the DOS varies with energy, leading to the formation of distinct energy bands and gaps.

$$g(E)=\left(\frac{8\pi\sqrt{2}}{h^3}\right)m^{3/2}\sqrt{E-E_c}$$

where $(g(E))$ is the density of states, (m) is the electron mass, (h) is Planck's constant, and (E_c) is the conduction band edge. Understanding the band theory of solids is crucial for the advancement of electrical materials and devices. This concept elucidates the reasons for the conductivity of some materials and the lack thereof in others, serving as the fundamental basis for contemporary semiconductor technology (Naumov & Hemley, 2015).

5.4. MAGNETISM AND SUPERCONDUCTIVITY

Magnetism and superconductivity are fundamental phenomena in the field of solid-state with wide-ranging applications in many technological and industrial sectors. Magnetism is the result of the magnetic moments of electrons in materials, which give rise to various magnetic properties such as ferromagnetism, antiferromagnetism, and paramagnetism. These characteristics are essential in the design of magnetic storage devices, sensors, and electric motors. Superconductivity, first observed in 1911, refers to the absence of electrical resistance in certain materials when they are cooled to very low temperatures. This phenomenon facilitates the development of very efficient power transmission networks and robust electromagnets used in MRI equipment and particle accelerators. The interaction between these two characteristics results in captivating materials and technologies, transforming several disciplines like medical imaging and quantum computing (Naito et al., 2016).

5.4.1. Magnetism

Magnetism is an inherent properties of materials that arises from the movement of electric charges. To understand magnetism, one must analyze magnetic moments and magnetic fields, which play a vital role in a wide range of applications, including data storage and medical imaging.

5.4.1.1. Magnetic Moments and Magnetic Fields

A magnetic moment $\left(\left(\vec{\mu}\right)\right)$ is a vector quantity that represents the magnetic strength and orientation of a magnet or other object that produces a magnetic field. In atoms, the magnetic moment arises from two primary sources: the spin of the electrons and their orbital motion around the nucleus.

The magnetic field $\left(\left(\vec{B}\right)\right)$ generated by a magnetic moment can be described using the Biot-Savart law for a current loop:

$$\vec{B} = \frac{\mu_0}{4\pi} \frac{I\, d\vec{l} \times \hat{r}}{r^2}$$

Where, (μ_0) is the permeability of free space $((4\pi\times10^{-7}\,T\cdot m/A));(I)$ is the current; $(d\vec{l})$ is the differential element of the wire; $(\hat{r})$ is the unit vector from the current element to the point where the field is calculated; and (r) is the distance from the current element to the point.

The total magnetic field generated by a magnetic dipole moment at a point $(\vec{r})$ is given by:

$$\vec{B}=\frac{\mu_0}{4\pi}\left(\frac{3(\vec{\mu}\cdot\hat{r})\hat{r}-\vec{\mu}}{r^3}\right)$$

Magnetic moments interact with external magnetic fields, resulting in forces and torques that align the moments with the direction of the field. This interaction serves as the foundation for the magnetic characteristics that are seen in materials (Ferrari et al., 1997).

5.4.1.2. Types of Magnetic Materials

Magnetic materials can be classified based on their response to an external magnetic field. The main types include:

1. Diamagnetic Materials:
   a. These materials lack a persistent magnetic moment. When subjected to an external magnetic field, they develop a weak, transient magnetic moment that opposes the applied field.

   b. Susceptibility $((\chi_m))$ is negative and very small $((\chi_m \approx -10^{-5}))$.
   c. Examples include bismuth, copper, and water.
2. Paramagnetic Materials:
   a. These materials generate unpaired electrons that generate persistent magnetic moments. When exposed to an external magnetic field, these moments align with the direction of the field, leading to a total positive magnetization.

   b. Susceptibility $((\chi_m))$ is positive but small $((\chi_m \approx 10^{-3}))$.
   c. Examples include aluminum, platinum, and oxygen (Koppe & Pell, 2023).
3. Ferromagnetic Materials:
   a. These materials possess unpaired electrons and exhibit a strong tendency for magnetic moments to align in parallel with each other within certain areas known as domains. Upon the application of an external magnetic

field, the domains undergo alignment, leading to a significant overall magnetization.

b. Susceptibility $((\chi_m))$ is very large and positive.
c. Examples include iron, cobalt, and nickel.

4. Antiferromagnetic Materials:
   a. These materials exhibit a phenomenon where their magnetic moments align in an alternating pattern, causing them to mutually nullify each other, resulting in the absence of any overall macroscopic magnetization.
   b. Examples include manganese oxide (MnO) and iron oxide (FeO).
5. Ferrimagnetic Materials:
   a. These materials have magnetic moments that align in an alternating pattern like antiferromagnets, but the opposing moments are unequal, resulting in a net magnetization.
   b. Examples include magnetite $(\mathrm{Fe}(_3)\mathrm{O}(_4)$ and yttrium iron garnet (YIG).

Understanding these classifications and their characteristics is essential for creating magnetic gadgets and materials for diverse technological applications. Magnetic materials are classified according to their atomic arrangement and the interactions between magnetic moments, which ultimately define their macroscopic magnetic characteristics (Bozorth & Kramer, 1959).

5.4.2. Magnetic Ordering

Magnetic ordering pertains to the configuration of magnetic moments in materials, dictating their magnetic characteristics. Understanding various forms of magnetic ordering is crucial in many technical applications, ranging from data storage to magnetic resonance imaging (Roy, 2013).

5.4.2.1. Ferromagnetism

Ferromagnetic materials display inherent magnetism below a certain temperature known as the Curie temperature (T_C). In these materials, adjacent atomic magnetic moments align in the same direction inside areas known as magnetic domains. The alignment remains unchanged even in the absence of an external magnetic field owing to the robust exchange interactions between neighboring magnetic moments.

5.4.2.2. Antiferromagnetism

In antiferromagnetic materials, neighboring atomic magnetic moments align antiparallel to each other, resulting in a net zero macroscopic magnetization. Antiferromagnets exhibit a Néel temperature (T_n) below which the antiparallel alignment occurs. Despite having no net magnetization, antiferromagnets play a crucial role in spintronics and magnetic sensors due to their ability to switch between magnetic states (Sokoloff, 1970).

5.4.2.3. Ferrimagnetism

Ferrimagnetic materials exhibit magnetic moments attraction that align similarly to antiferromagnets, but with varying strengths, leading to a resultant magnetization. The net magnetization is a result of the existence of two sublattices with distinct magnetic moments. Ferrimagnets play a crucial role in several applications, including magnetic recording medium and magnetic nanoparticles used in biological applications.

5.4.2.4. Magnetic Domains and Hysteresis

- **Magnetic Domains:** Magnetic domains are localized regions within a magnetic material where the magnetic moments of atoms are oriented in the same direction. Domains optimize the amount of energy of the material by facilitating the alignment of magnetic moments in a particular direction. The size and number of domains are contingent upon variables such as the morphology of the material, its crystalline arrangement, and the strength of the applied magnetic field. Understand and controlling the arrangement of domains is essential for enhancing the performance of magnetic materials in diverse applications.
- **Hysteresis:** Hysteresis occurs when a material's magnetization lags the magnetic field. The energy needed to reorient magnetic domains when the applied magnetic field changes causes hysteresis. Designing effective magnetic devices and materials requires minimizing hysteresis losses.

In conclusion, Magnetic ordering governs the magnetic characteristics and behaviors of materials. Ferromagnetism, antiferromagnetism, and ferrimagnetism determine magnetic moment alignment, affecting magnetic data storage, advanced magnetic sensors, and medical imaging (Chibowski, 2003).

5.5. SUPERCONDUCTIVITY

Superconductivity is an extraordinary phenomenon observed in certain materials, characterized by the absence of electrical resistance and the expulsion of magnetic fields when the temperature drops below a critical point. This property, in addition to the Meissner effect and the BCS theory, serves as the fundamental basis for a wide range of applications, including powerful electromagnets and high-speed digital circuits.

5.5.1. Meissner Effect

When a substance exhibits superconductivity, it forcefully removes magnetic fields from its interior. The expulsion is a result of the generation of supercurrents on the surface of the material, which effectively neutralize any magnetic flux that enters the substance. Consequently, when a superconductor is subjected to an external magnetic field, it will expel the magnetic field from its interior, displaying flawless diamagnetism. Walter Meissner and Robert Ochsenfeld were the first to discover this occurrence in 1933 (Xu et al., 1995).

5.5.2. Zero Resistance

One of the most renowned features of superconductors is their ability to transmit electric current without encountering any resistance. Superconducting materials below a critical temperature (T_C) exhibit the phenomenon of Cooper pairing, where electrons pair up and move collectively across the lattice without being disrupted by impurities or vibrations in the lattice. Superconductors, due to their lack of resistance, can maintain continuous currents indefinitely after being initiated.

5.5.3. BCS Theory of Superconductivity

The Bardeen-Cooper-Schrieffer (BCS) theory, formulated in 1957 by John Bardeen, Leon Cooper, and Robert Schrieffer, offers a thorough elucidation of superconductivity in typical superconductors. The BCS theory explains that superconductivity occurs when Cooper pairs are formed, which involves the interaction of two electrons with opposing spins via lattice vibrations known as phonons. The pair of electrons overcome the repulsion caused by Coulomb forces by reducing their combined energy, resulting in the formation of a condensate that demonstrates large-scale quantum coherence.

The key equations governing the BCS theory include:

1. Gap Equation:

$$\Delta(r) = -\frac{1}{V}\sum_{k} V(k)\frac{\Delta(k)}{2E_k}\tanh\tanh\left(\frac{E_k}{2k_BT}\right)$$

Remember

The Meissner Effect demonstrates the expulsion of magnetic fields from a superconductor when cooled below its critical temperature, highlighting its unique electromagnetic properties.

Here, $\Delta(r)$ represents the superconducting order parameter, $V(k)$ denotes the electron-phonon interaction potential, (E_k) is the energy of the Cooper pair, (k_B) is the Boltzmann constant, and (T) is the temperature.

2. BCS Gap Equation:

$$\Delta = \Delta_0 tanh\left(\frac{1.74k_BT_C}{\Delta_0}\right)$$

(Δ_0) is the superconducting energy gap at absolute zero temperature, and (T_C) is the critical temperature below which the material becomes superconducting.

The BCS theory effectively accounts for several experimental observations, including the temperature dependence of the superconducting transition, anomalies in specific heat, and the isotope effect. It has facilitated the understanding and advancement of other classes of superconductors, such as high-temperature superconductors and unconventional superconducting materials.

Superconductivity is a significant field of research in condensed matter physics, known for its characteristics such as the Meissner effect, zero resistance, and its explanation via the BCS theory (Johnston, 2013).

ACTIVITY 5.1.

Explore crystal structures using online tools or models to identify crystal planes and Miller indices, illustrating how these indices relate to the arrangement of atoms in a crystal lattice.

SUMMARY

This chapter provides a comprehensive overview of solid-state physics, beginning with the fundamental concepts of atomic bonding, including van der Waals attraction, repulsion, and the structure of various crystals, such as ionic crystals. The chapter delves into the role of quantum mechanics in bonding within solids, illustrating how it influences the physical properties of solids.

The discussion then moves to crystal structures and lattice dynamics, covering key topics such as Bravais lattices and the application of Miller indices to describe crystal planes. The chapter emphasizes the importance of understanding these structures for predicting and explaining the behavior of solid materials.

The electronic properties of solids are explored, with a focus on how quantum mechanics applies to these properties and the development of band theory. This section provides insight into how electrons behave within various types of solids, influencing their electrical and thermal properties.

Magnetism and superconductivity are also examined, with detailed explanations of magnetic ordering, the Meissner effect, and the BCS theory of superconductivity. These topics are crucial for understanding advanced materials and their technological applications.

REVIEW QUESTIONS

1. Explain the different types of atomic binding in solids, particularly focusing on van der Waals attraction and ionic bonding.
2. What are Bravais lattices, and why are they important in understanding crystal structures?
3. Describe the concept of Miller indices and their significance in identifying crystal planes.
4. Discuss the application of quantum mechanics in explaining the electronic properties of solids. How does it contribute to the development of band theory?
5. What is the Meissner effect, and how does it relate to the phenomenon of superconductivity?

MULTIPLE CHOICE QUESTIONS

1. Which of the following describes van der Waals attraction?

a. A type of ionic bonding between atoms

b. A weak intermolecular force due to induced dipole interactions

c. A strong covalent bond between molecules

d. An electrostatic attraction between ions

2. **What is a Bravais lattice?**
   a. A type of crystal defect
   b. A repeating arrangement of atoms in a crystal
   c. A method for measuring crystal planes
   d. A form of magnetic ordering in solids
3. **The Miller indices (hkl) are used to:**
   a. Determine the energy levels of electrons in a solid
   b. Identify and describe crystal planes
   c. Measure the strength of magnetic fields in materials
   d. Calculate the thermal conductivity of a material
4. **Band theory of solids is essential for understanding:**
   a. The thermal expansion of materials
   b. The electronic properties and conductivity of materials
   c. The optical properties of insulators
   d. The mechanical strength of polymers
5. **Which phenomenon is described by the Meissner effect?**
   a. The expulsion of magnetic fields from a superconductor
   b. The increase in electrical resistance at low temperatures
   c. The alignment of magnetic domains in a ferromagnet
   d. The generation of electric current in response to a magnetic field

Answers to Multiple Questions

1. (b); 2. (b); 3. (b); 4. (b); 5. (a)

REFERENCES

1. Abanin, D. A., & Pesin, D. A. (2011). Ordering of magnetic impurities and tunable electronic properties of topological insulators. *Physical Review Letters, 106*(13), 136802.
2. Ackland, G. J., Warren, M. C., & Clark, S. J. (1997). Practical methods in ab initio lattice dynamics. *Journal of Physics: Condensed Matter, 9*(37), 7861.
3. Andrews, L. C., & Bernstein, H. J. (1988). Lattices and reduced cells as points in 6-space and selection of Bravais lattice type by projections. *Acta Crystallographica Section A: Foundations of Crystallography, 44*(6), 1009–1018.
4. Aray, Y., Vidal, A. B., Rodriguez, J., Grillo, M. E., Vega, D., & Coll, D. S. (2009). First principles study of low Miller index RuS⊠ surfaces in hydrotreating conditions. *The Journal of Physical Chemistry C, 113*(45), 19545–19557.
5. Blakely, D. W., & Somorjai, G. A. (1977). The stability and structure of high Miller

index platinum crystal surfaces in vacuum and in the presence of adsorbed carbon and oxygen. *Surface Science, 65*(2), 419–442.

6. Bozorth, R. M., & Kramer, V. (1959). Some ferrimagnetic and antiferromagnetic materials at low temperatures. *Journal de Physique et le Radium, 20*(2, 3), 393–401.
7. Chamel, N. (2012). Neutron conduction in the inner crust of a neutron star in the framework of the band theory of solids. *Physical Review C—Nuclear Physics, 85*(3), 035801.
8. Chang, D. C., & Lee, Y. K. (2015). Study on the physical basis of wave-particle duality: Modelling the vacuum as a continuous mechanical medium. *Journal of Modern Physics, 6*(08), 1058.
9. Chibowski, E. (2003). Surface free energy of a solid from contact angle hysteresis. *Advances in Colloid and Interface Science, 103*(2), 149–172.
10. Choe, H. W., Kim, Y. J., Park, J. H., Morizumi, T., Pai, E. F., Krauss, N., & Ernst, O. P. (2011). Crystal structure of metarhodopsin II. *Nature, 471*(7340), 651–655.
11. Christianto, V. (2014). A review of Schrödinger equation and classical wave equation. *Prespacetime Journal, 5*(5), 587–590.
12. Donnay, J. H., & Harker, D. (1937). A new law of crystal morphology extending the law of Bravais. *American Mineralogist: Journal of Earth and Planetary Materials, 22*(5), 446–467.
13. Fagnocchi, S., Finazzi, S., Liberati, S., Kormos, M., & Trombettoni, A. (2010). Relativistic Bose–Einstein condensates: A new system for analogue models of gravity. *New Journal of Physics, 12*(9), 095012.
14. Ferrari, E. F., Da Silva, F. C. S., & Knobel, M. (1997). Influence of the distribution of magnetic moments on the magnetization and magnetoresistance in granular alloys. *Physical Review B, 56*(10), 6086.
15. Fianchini, M. (2017). Synthesis meets theory: Past, present and future of rational chemistry. *Physical Sciences Reviews, 2*(12), 20170134.
16. Gavezzotti, A. (1994). Are crystal structures predictable? *Accounts of Chemical Research, 27*(10), 309–314.
17. Gonze, X., Rignanese, G. M., & Caracas, R. (2005). First-principle studies of the lattice dynamics of crystals, and related properties. *Zeitschrift für Kristallographie—Crystalline Materials, 220*(5, 6), 458–472.
18. Guyer, R. A. (1970). The physics of quantum crystals. In *Solid State Physics* (Vol. 23, pp. 413–499). Academic Press.
19. Harder, M., & Hu, C. M. (2018). Cavity spintronics: An early review of recent progress in the study of magnon–photon level repulsion. *Solid State Physics, 69*, 47–121.
20. Harper, P. G., Hodby, J. W., & Stradling, R. A. (1973). Electrons and optic phonons in solids—the effects of longitudinal optical lattice vibrations on the electronic excitations of solids. *Reports on Progress in Physics, 36*(1), 1.
21. Heiss, W. D. (2000). Repulsion of resonance states and exceptional points. *Physical Review E, 61*(1), 929.

22. Herman, F. (1958). Theoretical investigation of the electronic energy band structure of solids. *Reviews of Modern Physics, 30*(1), 102.
23. Herrmann, W. A. (1986). Multiple bonds between transition metals and "bare" main group elements: Links between inorganic solid state chemistry and organometallic chemistry. *Angewandte Chemie International Edition in English, 25*(1), 56–76.
24. Hoffmann, R. (1987). How chemistry and physics meet in the solid state. *Angewandte Chemie International Edition in English, 26*(9), 846–878.
25. Ibach, H., & Lüth, H. (2013). *Solid-State Physics: An Introduction to Principles of Materials Science* (Vol. 1, pp. 65–69). Springer Science & Business Media.
26. Johnston, D. C. (2013). Elaboration of the α-model derived from the BCS theory of superconductivity. *Superconductor Science and Technology, 26*(11), 115011.
27. Jones, W., & March, N. H. (1985). *Theoretical Solid State Physics* (Vol. 35, pp. 34–40). Courier Corporation.
28. Koppe, J., & Pell, A. J. (2023). Structure determination and refinement of paramagnetic materials by solid-state NMR. *ACS Physical Chemistry Au, 3*(5), 419–433.
29. Kubota, K., Kumakura, S., Yoda, Y., Kuroki, K., & Komaba, S. (2018). Electrochemistry and solid-state chemistry of NaMeO⊠ (Me = 3d transition metals). *Advanced Energy Materials, 8*(17), 1703415.
30. Lazebnik, M., Okoniewski, M., Booske, J. H., & Hagness, S. C. (2007). Highly accurate Debye models for normal and malignant breast tissue dielectric properties at microwave frequencies. *IEEE Microwave and Wireless Components Letters, 17*(12), 822–824.
31. Madelung, O. (2012). *Introduction to Solid-State Theory* (Vol. 2, pp. 74–80). Springer Science & Business Media.
32. McLachlan, A. D. (1964). Van der Waals forces between an atom and a surface. *Molecular Physics, 7*(4), 381–388.
33. Moeck, P., & DeStefano, P. (2018). Accurate lattice parameters from 2D-periodic images for subsequent Bravais lattice type assignments. *Advanced Structural and Chemical Imaging, 4*, 1–33.
34. Morales, M. A., Clay, R., Pierleoni, C., & Ceperley, D. M. (2013). First principles methods: A perspective from quantum Monte Carlo. *Entropy, 16*(1), 287–321.
35. Müller, P. (2009). Practical suggestions for better crystal structures. *Crystallography Reviews, 15*(1), 57–83.
36. Naito, M., Krockenberger, Y., Ikeda, A., & Yamamoto, H. (2016). Reassessment of the electronic state, magnetism, and superconductivity in high-Tc cuprates with the Nd⊠CuO⊠ structure. *Physica C: Superconductivity and its Applications, 523*, 28–54.
37. Naumov, I. I., & Hemley, R. J. (2015). Origin of transitions between metallic and insulating states in simple metals. *Physical Review Letters, 114*(15), 156403.
38. Odobel, F., & Pellegrin, Y. (2013). Recent advances in the sensitization of wide-band-gap nanostructured p-type semiconductors: Photovoltaic and photocatalytic applications. *The Journal of Physical Chemistry Letters, 4*(15), 2551–2564.

39. Ono, Y., Ochi, T., Abe, Y., & Hasegawa, A. (2005). Classification of photonic crystals with Bravais lattice structure fabricated by four-beam interference. In *Nanoengineering: Fabrication, Properties, Optics, and Devices II*. SPIE. (Vol. 5931, pp. 13–20).

40. Panich, A. M. (2008). Electronic properties and phase transitions in low-dimensional semiconductors. *Journal of Physics: Condensed Matter, 20*(29), 293202.

41. Patsalas, P., Kalfagiannis, N., Kassavetis, S., Abadias, G., Bellas, D. V., Lekka, C., & Lidorikis, E. (2018). Conductive nitrides: Growth principles, optical and electronic properties, and their perspectives in photonics and plasmonics. *Materials Science and Engineering: R: Reports, 123*, 1–55.

42. Paulus, E. F., Leusen, F. J., & Schmidt, M. U. (2007). Crystal structures of quinacridones. *CrystEngComm, 9*(2), 131–143.

43. Perdew, J. P., Yang, W., Burke, K., Yang, Z., Gross, E. K., Scheffler, M., & Görling, A. (2017). Understanding band gaps of solids in generalized Kohn–Sham theory. *Proceedings of the National Academy of Sciences, 114*(11), 2801–2806.

44. Pitteri, M., & Zanzotto, G. (1996). On the definition and classification of Bravais lattices. *Acta Crystallographica Section A: Foundations of Crystallography, 52*(6), 830–838.

45. Pohl, H. A. (1964). Physico-chemical aspects of organic semiconductors. *Progress in Solid State Chemistry, 1*, 316–346.

46. Raty, J. Y., Schumacher, M., Golub, P., Deringer, V. L., Gatti, C., & Wuttig, M. (2019). A quantum-mechanical map for bonding and properties in solids. *Advanced Materials, 31*(3), 1806280.

47. Roy, S. B. (2013). First order magneto-structural phase transition and associated multi-functional properties in magnetic solids. *Journal of Physics: Condensed Matter, 25*(18), 183201.

48. Sangian, H. F., Aruan, I., Tamuntuan, G. H., Bobanto, M. D., Sadjab, B. A., Purwadi, R., & Gunawan, S. (2019). A comparative study of the structures, crystallinities, Miller indices, crystal parameters, and particle sizes of microwave- and saline-water-treated cassava starch. *BioResources, 14*(4), 8212–8228.

49. Schwarz, R. B., & Johnson, W. L. (1983). Formation of an amorphous alloy by solid-state reaction of the pure polycrystalline metals. *Physical Review Letters, 51*(5), 415.

50. Sokoloff, J. B. (1970). Antiferromagnetism in narrow-band solids. *Physical Review B, 1*(3), 1144.

51. Sólyom, J. (2008). *Fundamentals of the Physics of Solids: Volume II: Electronic Properties* (Vol. 2, pp. 98–100). Springer Science & Business Media.

52. Stuke, J. (1970). Review of optical and electrical properties of amorphous semiconductors. *Journal of Non-Crystalline Solids, 4*, 1–26.

53. Sun, S., Zhang, X., Cui, J., & Liang, S. (2020). Identification of the Miller indices of a crystallographic plane: A tutorial and a comprehensive review on fundamental theory, universal methods based on different case studies, and matters needing attention. *Nanoscale, 12*(32), 16657–16677.

54. Tasker, P. W. (1979). The stability of ionic crystal surfaces. *Journal of Physics C: Solid State Physics, 12*(22), 4977.

55. Tian, N., Zhou, Z. Y., & Sun, S. G. (2008). Platinum metal catalysts of high-index surfaces: From single-crystal planes to electrochemically shape-controlled nanoparticles. *The Journal of Physical Chemistry C, 112*(50), 19801–19817.

56. Weiss, R. J. (2013). *Solid State Physics for Metallurgists: International Series of Monographs on Metal Physics and Physical Metallurgy* (Vol. 6, pp. 69–75). Elsevier.

57. Xu, D., Yip, S. K., & Sauls, J. A. (1995). Nonlinear Meissner effect in unconventional superconductors. *Physical Review B, 51*(22), 16233.

58. Yang, H., Kim, S. W., Chhowalla, M., & Lee, Y. H. (2017). Structural and quantum-state phase transitions in van der Waals layered materials. *Nature Physics, 13*(10), 931–937.

59. Zhong, X., & Zhao, Y. (2013). Non-Markovian stochastic Schrödinger equation at finite temperatures for charge carrier dynamics in organic crystals. *The Journal of Chemical Physics, 138*(1), 87–90.

CHAPTER

6 Optics and Photonics

LEARNING OBJECTIVES

After studying this chapter, you will be able to:

- Understand the principles of wave optics, including the phenomena of interference and diffraction.
- Describe the basic concepts of quantum optics and the nature of photons and quantum states of light.
- Explain the principles of laser operation, including spontaneous and stimulated emission.
- Analyze the concept of population inversion and its role in achieving optical gain in lasers.
- Explore the functioning and applications of photonic devices such as LEDs and fiber optic communication systems.
- Apply the principles of optics and photonics to modern communication technologies.
- Examine the technological advancements and applications of laser technology in various fields.

KEY TERMS FROM THIS CHAPTER

Diffraction	Fiber optic communication
Interference	Laser operation
Optical gain	Photons
Population inversion	Quantum optics
Spontaneous emission	Stimulated emission

UNIT INTRODUCTION

Wave optics, or physical optics, is the field of optics that examines the characteristics of light as a wave. In contrast to geometrical optics, wave optics explains phenomena like interference and diffraction that cannot be adequately described by ray theory alone. Interference arises when multiple light waves intersect, resulting in the creation of a pattern consisting of bright and dark fringes caused by constructive and destructive interference. Diffraction is the phenomenon where light waves bend and spread over barriers and through slits, creating distinct patterns that demonstrate the wave-like properties of light (Smith & King, 2001).

Wave optics marks a departure from geometric optics, offering insights into how light waves propagate, interact, and diffract when encountering various obstacles and openings. Interference, explored through experiments like Young's double-slit experiment, showcases how coherent light sources generate distinct patterns of constructive and destructive interference, essential in fields ranging from astronomy to quantum mechanics. Diffraction, on the other hand, demonstrates how light bends around edges and spreads out through narrow openings, exemplified by Huygens' principle and the intricate patterns produced by diffraction gratings used in optical instruments and spectroscopy (Lebeau and Innocenzi, 2011).

Moving into quantum optics and laser technology, the chapter explores the quantum mechanical nature of light. It explains the wave-particle duality of photons and examines how principles such as stimulated emission and population inversion underpin the operation of lasers. These technologies have not only revolutionized fields like medicine and telecommunications but also paved the way for innovations in quantum computing and high-precision measurements. Additionally, the chapter explores the practical applications of photonic devices, such as light-emitting diodes (LEDs) and fiber optic communication systems, highlighting their roles in modern displays, telecommunications networks, and scientific instrumentation (Miri and Alu, 2019).

6.1. WAVE OPTICS

Wave optics, also known as physical optics, deals with the study of light as a wave phenomenon. This field is essential for explaining various optical effects that cannot be accounted for by geometrical optics, which treats light as a collection of rays. The fundamental principles of wave optics include the superposition principle, which states that when two or more waves overlap, the resultant displacement at any point is the sum of the displacements due to each wave. This principle underlies the phenomena of interference and diffraction (Xiao & Voelz, 2006).

Wave optics differs significantly from geometrical optics. While geometrical optics uses the laws of reflection and refraction to describe light propagation, it fails to explain phenomena like interference and diffraction. These phenomena arise due to the wave nature of light, which can be described by the wave equation (Shimobaba et al., 2012):

$$\nabla^2 E - \frac{1}{c^2}\frac{\partial^2 E}{\partial t^2} = 0$$

where (E) is the electric field, (∇^2) is the Laplacian operator, and (c) is the speed of light. In wave optics, Huygens' principle plays a crucial role, stating that every point on a wavefront can be considered a source of secondary wavelets that spread out in all directions. The new wavefront is the envelope of these secondary wavelets, providing a basis for understanding diffraction and the propagation of light waves (Lax et al., 1975).

6.1.1. Interference of Light

Light interference is a phenomenon that happens when two or more coherent light waves overlap, creating a pattern of alternating bright and dark fringes. This phenomenon is a direct result of the wave-like properties of light and is fundamental to the study of wave optics (Moreira et al., 1996).

6.1.1.1. Conditions for Interference

For interference to occur, two primary requirements must be met:

1. **Coherence:** The light sources must exhibit coherence, meaning they maintain a consistent phase difference. Typically, this is accomplished by dividing a single light source into two or more beams.
2. **Path Difference:** The light waves must have a consistent difference in their paths. The disparity along the path determines whether the interference will result in constructive or destructive interference (Wolf, 1954).

6.1.1.2. Young's Double-Slit Experiment

Young's double-slit experiment is widely recognized as one of the most prominent demonstrations of interference. For this experiment, a laser or another type of coherent light source is used to shine light onto a screen that has two slits positioned close each other. The light that passes through these holes behaves as two coherent sources that create interference with each other (Liu et al., 2006).

Did you know?

Interference of light demonstrates how waves combine to form patterns, highlighting phenomena like constructive and destructive interference, essential in optics and photonics.

The resultant interference pattern observed on a distant screen comprises bright and dark fringes. The bright fringes, which result from constructive interference, arise when the difference in the paths traveled by the two waves is a integer multiple of the wavelength (λ):

$$dsinsin\,\theta = m\lambda$$

where (d) represents the separation between the slits, (θ) denotes the angle formed by the fringe with respect to the central axis, (m) signifies the order of the fringe (an integer), and (λ) denotes the wavelength of the light (Figure 6.1).

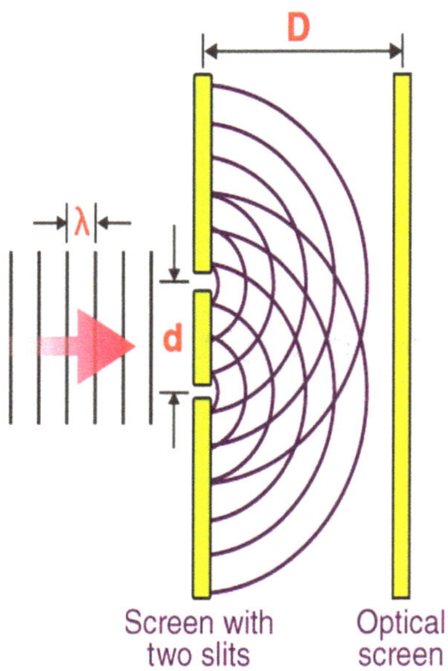

Figure 6.1. *Young's double slit experiment.*

Source: BYJU's, Creative Commons License.

The dark fringes, which result from destructive interference, occur when the difference in the paths traveled by the waves is an odd multiple of half the wavelength (Peng, 2022):

$$dsinsin\,\theta = \left(m + \frac{1}{2}\right)\lambda$$

6.1.1.3. Constructive and Destructive Interference

- **Constructive Interference:** This occurs when the phase difference between the two interfering waves results in their amplitudes adding up, resulting in a brighter fringe. Mathematically, it is expressed as:

$$\Delta\phi = 2n\pi$$

where $(\Delta\phi)$ is the phase difference, and (n) is an integer.

- **Destructive Interference:** This occurs when the phase difference causes the waves to cancel each other out, resulting in a dark fringe. This condition is described by (Zhang et al., 2020) (Figure 6.2):

$$\Delta\phi = (2n+1)\pi$$

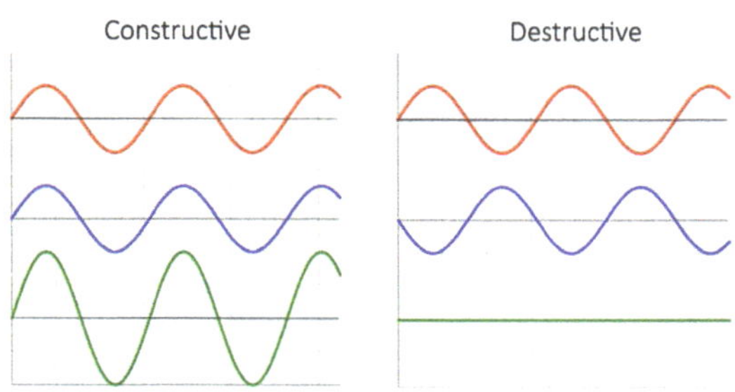

Figure 6.2. *Illustration of constructive and destructive interference.*

Source: Dina Zhabinskaya, Creative Commons License.

6.1.1.3.1. Applications of Interference

Interference is not only a fundamental concept in physics but also has practical applications. Some key applications include:

1. **Thin Film Interference:** Used in anti-reflective coatings on lenses and optical instruments to reduce glare and enhance transmission.
2. **Interferometry:** Techniques such as Michelson and Mach-Zehnder interferometers are employed for precise measurements of distance, surface irregularities, and refractive index changes.
3. **Holography:** A method for recording and reconstructing three-dimensional images by capturing the interference pattern of light waves reflected from an object.
6. **Fiber Optic Sensors:** Used in various sensing applications, including temperature, pressure, and strain measurement, where interference patterns provide precise information about environmental changes.

Interference of light reveals the wave properties of light and has significant technological implications, from enhancing optical devices to enabling precise measurements in scientific research (Dattner and Yadid-Pecht, 2011).

6.1.2. Diffraction of Light

Diffraction of light is the phenomenon where light waves bend and spread out when they encounter an object or pass through a small opening. The explanation for this phenomenon lies in the wave nature of light and is crucial for understanding different optical effects and technologies (Akhmanov et al., 1968).

6.1.2.1. Huygens' Principle

Huygens' Principle is a key principle in the field of wave optics which asserts that each point on a wavefront function as a source of secondary spherical wavelets. The new wavefront is formed by the outer boundary of these secondary wavelets. This theory aids understanding of the behavior of light waves as they come across obstructions or apertures. Huygens' Principle can be employed mathematically to derive the wave equation and elucidate the transmission of light waves:

$$\Psi(r,t)={}_{S}\frac{e^{ikr}}{r}\left(\frac{\partial\Psi}{\partial n}-ik\Psi\right)dS$$

where (Ψ) is the wave function, (k) is the wave number, (r) is the distance, and (S) is the surface over which the integration is performed (Romero and Hernández, 2008).

6.1.2.2. Single-Slit and Double-Slit Diffraction

Single-slit diffraction refers to the phenomenon that occurs when light passes through a small slit, causing it to spread out and form a pattern of alternating dark and bright fringes on a screen. The intensity pattern is determined by:

$$II(\theta)=I_0\left(\frac{sinsin(\beta)}{\beta}\right)^2$$

where $\left(\beta=\frac{\pi asinsin\,\theta}{\lambda}\right)$, (a) is the slit width, (θ) is the angle of diffraction, and (λ) is the wavelength of light.

Double-Slit Diffraction is a phenomenon that results from the combination of interference and diffraction phenomena. When light passes through two closely spaced apertures, each aperture undergoes diffraction, and the resultant waveforms interfere, generating a pattern of dark and bright bands. The resulting intensity pattern is determined by a combination of factors:

$$I(\theta)=I_0\left(coscos\left(\frac{\pi dsinsin\,\theta}{\lambda}\right)\right)^2\left(\frac{sinsin(\beta)}{\beta}\right)^2$$

where (d) is the distance between the slits (Dutt et al., 2021).

6.1.2.3. Diffraction Gratings

Diffraction Grating is composed of numerous evenly spaced parallel slits or grooves. When light passes through or reflects off the grating, it undergoes diffraction and interference, resulting in the formation of distinct and clearly defined spectral lines. The grating equation provides the condition for constructive interference in a diffraction grating:

$$dsinsin\,\theta=m\lambda$$

where (d) is the grating spacing, (θ) is the diffraction angle, (m) is the order of the diffraction, and (λ) is the wavelength of light (Figure 6.3).

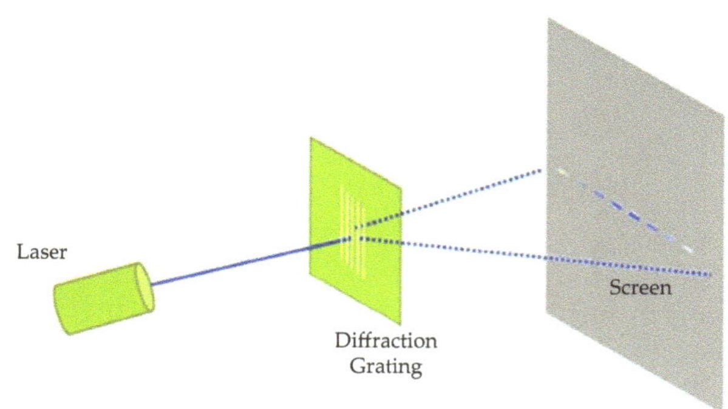

Figure 6.3. *Setup with diffraction grating.*

Source: Go Photonics, Creative Commons License.

Diffraction gratings are widely used in spectroscopy to disperse light into its component wavelengths for detailed analysis.

6.1.2.4. Applications of Diffraction

Diffraction has numerous practical applications in science and technology:

1. Spectroscopy: Diffraction gratings are essential components of spectrometers used to analyze the spectral composition of light from various sources, including stars and chemical compounds.
2. Optical Instruments: Understanding diffraction limits is crucial for designing high-resolution microscopes and telescopes.
3. CDs and DVDs: Data storage on CDs and DVDs relies on the diffraction of laser light from the micro-grooves on the disc surface.
4. X-ray Crystallography is a scientific method that involves analyzing the way X-rays are diffracted by the regular arrangement of atoms in a crystal lattice to identify the precise atomic structure of materials.

Diffraction reveals the wave nature of light and provides insights into the design and functionality of various optical devices and technologies (Pachon et al., 2016).

6.2. QUANTUM OPTICS AND LASER TECHNOLOGY

Quantum optics explores the interaction of light with matter at the quantum level, revealing phenomena that classical optics cannot explain, such as photon entanglement and quantum coherence. This field has led to groundbreaking advancements in laser technology, where principles of stimulated emission and quantum mechanics enable the generation of highly coherent and monochromatic light. Lasers are indispensable in modern technology, with applications ranging from telecommunications and medical diagnostics to cutting-edge research in quantum computing and encryption. Together, quantum optics and laser technology together form the backbone of many technological innovations and scientific discoveries (Chow & Jahnke, 2013).

6.2.1. Introduction to Quantum Optics

Quantum optics is a branch of physics that explores the interaction of light with matter at the quantum level. This field leverages the principles of quantum mechanics to describe the properties and behavior of photons, the fundamental particles of light. Understanding these interactions is crucial for advancing various technologies, such as lasers and quantum communication systems.

6.2.1.1. Basics of Quantum Optics

Quantum optics explores phenomena that are beyond the scope of classical optics, such as quantum entanglement and quantum coherence. Quantum electrodynamics (QED) is used to explain the interaction between light and matter, with a particular focus on the particle-like behavior of light, manifested through photons.

$$E = h\nu$$

$where(E)$ is the energy of a photon, (h) is Planck's constant, and (ν) is the frequency of the light (Fabre, 1999).

6.2.1.2. Wave-Particle Duality of Light

The phenomenon of wave-particle duality, which refers to the ability of light to exhibit both wave-like and particle-like properties, is a key principle in the field of quantum optics. Light demonstrate both wave-like and particle-like characteristics, depending on the specific experimental arrangement.

- **Wave Nature:** According to Maxwell's equations, light exhibits the properties of an electromagnetic wave, which may be defined by its wavelength (λ) and frequency (v).

$$c = \lambda \nu$$

where (c) is the speed of light.

- **Particle Nature:** In specific studies, light exhibits the characteristics of particles known as photons. The duality is evident in the double-slit experiment, where light exhibits both wave-like behavior by producing an interference pattern and particle-like behavior by being recognized as individual photons.

$$p = \frac{h}{\lambda}$$

where (p) is the momentum of a photon.

Quantum optics integrates these two features to provide a comprehensive description of interactions between light and matter. For instance, the photoelectric effect serves as evidence for the particle nature of light, as photons dislodge electrons from the surface of a substance.

$$E = h\nu - \phi$$

where (ϕ) represents the work function of the material. The equation demonstrates that the energy of incoming photons (hv) must exceed the work function to liberate electrons. Quantum optics utilizes these concepts to elucidate various optical phenomena and serves as the foundation for the technologies employed in devices such as lasers, which depend on the quantum mechanical phenomenon of stimulated emission (Hendry, 1980).

6.2.2. Photons and Quantum States of Light

Photons are the fundamental particles of light, and their properties are central to understanding quantum optics. This section explores how photons are quantized and the various quantum states of light that arise from their interactions.

6.2.2.1. Photons and Quantization of Light

Quantization of light refers to the discrete nature of photons, which are packets of electromagnetic energy. Each photon carries energy $(E = h\nu)$, where (h) is Planck's constant and (ν) is the frequency of the light. This quantization is crucial in explaining the photoelectric effect and other phenomena where light behaves as particles (Lupton & Vogelsang, 2021).

$$E = h\nu$$

6.2.2.2. Quantum States: Coherent, Fock, and Squeezed States

- **Quantum States:** Are of light describe different distributions of photons and their quantum mechanical properties. These states play a significant role in applications such as quantum information processing and quantum communication.
- **Coherent States:** They are classical-like states of light in which photons have well-defined phases and a Poissonian distribution of photon numbers. These states are essential in laser physics and interferometry.

$$\left| \alpha = e^{-|\alpha|^2/2} \sum_{n=0}^{\infty} \frac{\alpha^n}{\sqrt{n!}} \right| n$$

where $|\alpha|$ is the coherent state, (α) is a complex number representing the mean amplitude, and $|n|$ denotes the number state with (n) photons.

- **Fock States:** Fock states describe states with a fixed number of photons (n). These states are quantized and are used to model photon statistics in experiments.

$$\left| n = \frac{\left(a^{\dagger}\right)^{n}}{\sqrt{n!}} \right| 0$$

where $(a^{\dagger})$ is the photon creation operator and $(|0|)$ is the vacuum state.

- **Squeezed States:** Squeezed states exhibit reduced quantum noise in one quadrature component of the electromagnetic field at the expense of increased noise in the conjugate component. They find applications in quantum metrology and quantum computing.

$$|\zeta = \hat{S}(\zeta)|0$$

where $\hat{S}(\zeta)$ is the squeezing operator and (ζ) is the squeezing parameter.

Understanding these quantum states is crucial for manipulating and harnessing light at the quantum level, paving the way for advancements in fields such as quantum cryptography, quantum sensing, and quantum computing (Celeghini et al., 1995).

Remember

Understanding the principles of laser operation involves knowing how stimulated emission amplifies light into a coherent beam, crucial for applications in various fields.

6.2.3. Principles of Laser Operation

Lasers (Light Amplification by Stimulated Emission of Radiation) operate based on fundamental principles of quantum optics and electromagnetic theory.

This section delves into the processes that enable laser operation, including spontaneous and stimulated emission, population inversion, and the different types of lasers used in various applications.

6.2.4. Spontaneous and Stimulated Emission

- **Spontaneous Emission:** This refers to the process in which an atom or molecule in an excited state transitions to a lower energy state without any external influence, resulting in the emission of a photon.

$$\gamma_A = A_{21}\rho(\nu)$$

The symbol (γ_A) represents the coefficient for spontaneous emission, (A_{21}) represents the Einstein coefficient for spontaneous emission, and ρ(v) represents the photon density of states.

- **Stimulated Emission:** This is the process in which an incoming photon interacts with an atom or molecule that is already in an excited state. This interaction causes the atom or molecule to transition to a lower energy state and release a photon with the same frequency, phase, and direction as the incoming photon.

$$\gamma_S = B_{21}\rho(\nu)I(\nu)$$

where (γ_S) is the stimulated emission coefficient, (B_{21}) is the Einstein coefficient for stimulated emission, $\rho(\nu)$ is the photon density of states, and $I(\nu)$ is the intensity of the incident radiation (Friedman et al., 1988).

6.2.5. Population Inversion and Optical Gain

- **Population Inversion:** In laser operation, achieving population inversion involves having a higher population of atoms or molecules in excited states than in the ground state. This is essential for laser amplification.

$$N_2 > N_1$$

where (N_2) is the population of atoms or molecules in the excited state, and (N_1) is the population in the ground state.

- **Optical Gain:** Optical gain refers to the amplification of light signals within the laser medium due to stimulated emission dominating absorption losses.

$$G = \frac{N_2}{N_1}$$

where (G) is the optical gain, (N_2) is the population of excited states, and (N_1) is the population of ground states (Frogley et al., 2006).

6.3. PHOTONIC DEVICES

Photonic devices harness the principles of optics and photonics to manipulate and control light for various applications across diverse fields. These devices are pivotal in modern technology, enabling advancements in telecommunications, information processing, sensing, and medical diagnostics. Optical fibers are a prime example of a technology that has transformed communication by efficiently delivering data over significant distances. Photonic sensors use light to accurately and sensitively detect and quantify factors such as temperature, pressure, and chemical compositions. Furthermore, photonic integrated circuits (PICs) combine numerous optical components onto a single chip, enabling the creation of small and efficient devices for data communication and processing. Photonics is advancing rapidly, and these devices provide new possibilities in areas such as renewable energy and quantum computing, highlighting their crucial role in defining the future of technology (Jiang et al., 2021).

6.3.1. Light Emitting Diodes (LEDs)

LEDs function through electroluminescence, a process in which the recombination of electrons and electron holes in a semiconductor material leads to the production of light. The process begins by injecting electrons and holes into the semiconductor junction while applying a forward bias. Electrons in the conduction band combine with holes in the valence band, emitting energy as photons. The energy of the emitted photons is directly proportional to the bandgap energy of the semiconductor material (Figure 6.4).

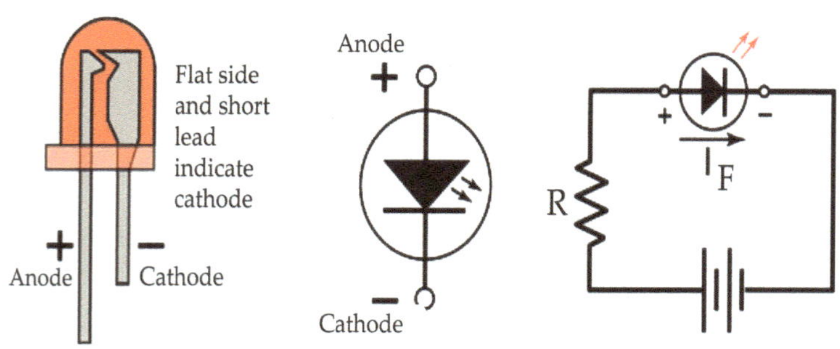

Figure 6.4. *Illustration of light emitting diodes (LEDs).*

Source: Hyper Physics, Creative Commons License.

6.3.1.1. Material Properties and Efficiency

LEDs are predominantly manufactured using semiconductor materials such as gallium arsenide (GaAs), gallium nitride (GaN), or indium gallium nitride (InGaN), depending on the desired wavelength of light emitted. The efficiency of light-emitting diodes (LEDs) is determined by their external quantum efficiency (EQE) and internal quantum efficiency (IQE). EQE quantifies the proportion of photons emitted compared to the number of electrons injected, whereas IQE evaluates the percentage of recombination events that lead to the emission of photons. Maximizing LED efficiency requires improving material quality, optimizing device design, and enhancing thermal management (Luo & Jen, 2013).

Tip

Utilizing LEDs in everyday electronics reduces energy consumption due to their efficiency in converting electricity to light, making them ideal for lighting and display applications.

6.3.1.2. Applications of LEDs in Lighting and Displays

LEDs have revolutionized lighting and display technologies due to their high efficiency, long lifespan, and compact size. In lighting applications, LEDs are used in residential, commercial, and industrial settings for general illumination, architectural lighting, and automotive lighting. They offer advantages such as instant switching, dimming capability, and directional lighting. In displays, LEDs are employed in flat panel displays, signage, and electronic screens, providing vibrant colors and high contrast ratios. Ongoing research aims to further enhance efficiency, expand color options, and reduce costs to broaden the scope of LED applications (Mottier, 2010).

6.3.2. Fiber Optic Communication

Fiber optic communication revolutionized telecommunications by utilizing the principles of total internal reflection to transmit data through optical fibers. These fibers, typically made of glass or plastic, guide light pulses over long distances with minimal loss of signal strength. The fundamental operation relies on the principle that light, when traveling through a medium with a lower refractive index than its surroundings, undergoes internal reflection upon encountering the interface, ensuring minimal dispersion and attenuation (Nyachionjeka et al., 2020).

6.3.2.1. Basics of Fiber Optics

Fiber optics operates on the principle of transmitting light through a core surrounded by a cladding layer with a lower refractive index, which ensures that light signals propagate through the fiber by total internal reflection. This core-cladding structure allows multiple light signals to be transmitted simultaneously, greatly enhancing bandwidth and data transfer rates compared to traditional copper cables.

6.3.2.2. Optical Fibers and Signal Transmission

Optical fibers are primarily composed of a core, through which light propagates, surrounded by a cladding layer that reflects light back into the core. This structure enables the transmission of light signals with minimal loss and interference, making fiber optics ideal for long-distance communication networks (Shi et al., 2020).

6.3.2.3. Applications in Telecommunications

Fiber optic communication finds widespread applications in telecommunications, including high-speed internet, telephone services, and cable television transmissions. Its ability to carry large amounts of data over significant distances with low attenuation has made it indispensable to modern communication infrastructure.

6.3.2.4. Advantages and Challenges of Fiber Optic Systems

6.3.2.4.1. Advantages

1. High data transmission rates;
2. Immunity to electromagnetic interference;
3. Minimal signal degradation over long distances.

6.3.2.4.2. Challenges

1. High initial installation costs;
2. Susceptibility to physical damage;
3. Need for specialized equipment for splicing and maintenance (Shtaif et al., 2022).

ACTIVITY 6.1.

Conduct a diffraction experiment using a laser pointer and a small aperture to observe diffraction patterns, reinforcing the principles of wave optics and light behavior.

SUMMARY

This chapter covers the fundamental principles of optics and photonics, starting with wave optics. It explains interference and diffraction of light, essential concepts for understanding various optical phenomena. The discussion then shifts to quantum optics, introducing the quantum nature of light and the concept of photons. The chapter also delves into the principles of laser operation, detailing the processes of spontaneous and stimulated emission, as well as the critical concept of population inversion which is necessary to achieve optical gain in lasers.

Furthermore, the chapter explores various photonic devices, including light-emitting diodes (LEDs) and their role in modern technology. The principles of fiber optic communication are also discussed, highlighting their importance in contemporary communication systems. Overall, this chapter provides a comprehensive understanding of the principles, technologies, and applications related to optics and photonics, emphasizing their significance in scientific and technological advancements.

REVIEW QUESTIONS

1. Explain the difference between interference and diffraction of light and provide examples of each phenomenon.
2. Describe the principles behind laser operation, including the processes of spontaneous and stimulated emission.
3. How does population inversion contribute to the production of laser light? Explain with a diagram if possible.
5. Discuss the applications of fiber optic communication and compare its advantages over traditional copper wire communication.
6. What are some of the key technological applications of LEDs in modern society?

MULTIPLE CHOICE QUESTIONS

1. Which optical phenomenon occurs when two or more light waves overlap and create regions of constructive and destructive interference?

a. Diffraction

b. Reflection

c. Interference

d. Refraction

2. What is the fundamental unit of light in quantum optics?

a. Photon

b. Electron

c. Proton

d. Neutron

3. **In laser technology, stimulated emission refers to:**
   a. The absorption of photons by excited atoms
   b. The emission of photons due to spontaneous decay
   c. The emission of additional photons by excited atoms
   d. The conversion of thermal energy into light energy
4. **Population inversion in lasers occurs when:**
   a. More atoms are in the ground state than in the excited state
   b. The number of excited atoms equals the number of ground state atoms
   c. More atoms are in the excited state than in the ground state
   d. The laser beam splits into multiple directions
5. **Fiber optic communication is preferred over copper wire communication mainly because:**
   a. It is less expensive
   b. It allows for faster data transmission
   c. It is easier to install
   d. It requires less maintenance

Answers to Multiple Questions

1. (c); 2. (a); 3. (c); 4. (c); 5. (b)

REFERENCES

1. Akhmanov, S. A., Sukhorukov, A. P., & Khokhlov, R. V. (1968). Self-focusing and diffraction of light in a nonlinear medium. *Soviet Physics Uspekhi, 10*(5), 609.
2. Celeghini, E., Demartino, S., Desiena, S., Rasetti, M., & Vitiello, G. (1995). Quantum groups, coherent states, squeezing and lattice quantum mechanics. *Annals of Physics, 241*(1), 50–67.
3. Chow, W. W., & Jahnke, F. (2013). On the physics of semiconductor quantum dots for applications in lasers and quantum optics. *Progress in Quantum Electronics, 37*(3), 109–184.
4. Dattner, Y., & Yadid-Pecht, O. (2011). Analysis of the effective refractive index of silicon waveguides through the constructive and destructive interference in a Mach–Zehnder interferometer. *IEEE Photonics Journal, 3*(6), 1123–1132.
5. Dutt, S., Dutt, N., & Dutt, A. (2021). Light strands: Visualization of free space in double slit diffraction. *International Journal of Physics, 9*(4), 197–205.
6. Fabre, C. (1999). Basics of quantum optics and cavity quantum electrodynamics. In *Confined Photon Systems: Fundamentals and Applications: Lectures from the Summerschool Held in Cargèse, Corsica, 3–15 August 1998* (Vol. 1, pp. 1–37). Springer Berlin Heidelberg.

7. Friedman, A., Gover, A., Kurizki, G., Ruschin, S., & Yariv, A. (1988). Spontaneous and stimulated emission from quasifree electrons. *Reviews of Modern Physics, 60*(2), 471.
8. Frogley, M. D., Dynes, J. F., Beck, M., Faist, J., & Phillips, C. C. (2006). Gain without inversion in semiconductor nanostructures. *Nature Materials, 5*(3), 175–178.
9. Hendry, J. (1980). The development of attitudes to the wave-particle duality of light and quantum theory, 1900–1920. *Annals of Science, 37*(1), 59–79.
10. Jiang, J., Chen, M., & Fan, J. A. (2021). Deep neural networks for the evaluation and design of photonic devices. *Nature Reviews Materials, 6*(8), 679–700.
11. Lax, M., Louisell, W. H., & McKnight, W. B. (1975). From Maxwell to paraxial wave optics. *Physical Review A, 11*(4), 1365.
12. Lebeau, B., & Innocenzi, P. (2011). Hybrid materials for optics and photonics. *Chemical Society Reviews, 40*(2), 886–906.
13. Liu, X. J., Cherepkov, N. A., Semenov, S. K., Kimberg, V., Gel'mukhanov, F., Prümper, G., & Ueda, K. (2006). Young's double-slit experiment using core-level photoemission from N2: Revisiting Cohen–Fano's two-centre interference phenomenon. *Journal of Physics B: Atomic, Molecular and Optical Physics, 39*(23), 4801.
14. Luo, J., & Jen, A. K. Y. (2013). Highly efficient organic electrooptic materials and their hybrid systems for advanced photonic devices. *IEEE Journal of Selected Topics in Quantum Electronics, 19*(6), 42–53.
15. Lupton, J. M., & Vogelsang, J. (2021). Photon correlations probe the quantized nature of light emission from optoelectronic materials. *Applied Physics Reviews, 8*(4), 258–260.
16. Miri, M. A., & Alu, A. (2019). Exceptional points in optics and photonics. *Science, 363*(6422), eaar7709.
17. Moreira, A. J. C., Valadas, R. T., & de Oliveira Duarte, A. M. (1996). Performance of infrared transmission systems under ambient light interference. *IEE Proceedings–Optoelectronics, 143*(6), 339–346.
18. Mottier, P. (2010). *LED for Lighting Applications* (Vol. 134, pp. 214–218). John Wiley & Sons.
19. Nyachionjeka, K., Tarus, H., & Langat, K. (2020). Design of a photonic crystal fiber for optical communications application. *Scientific African, 9*, 511.
20. Pachon, E. G. P., Zamboni-Rached, M., Dorrah, A. H., Mojahedi, M., Gesualdi, M. R. R., & Cabrera, G. G. (2016). Architecting new diffraction-resistant light structures and their possible applications in atom guidance. *Optics Express, 24*(22), 25403–25408.
21. Peng, H. (2022). Non-interference patterns evolving to interference patterns in double slit experiments and in cross double slit experiments. *Research Square*, preprint, DOI: https://doi.org/10.21203/rs.3.587-590.
22. Romero, J. A., & Hernández, L. (2008). Diffraction by a circular aperture: An application of the vectorial theory of Huygens's principle in the near field. *JOSA A, 25*(8), 2040–2043.

23. Shi, W., Tian, Y., & Gervais, A. (2020). Scaling capacity of fiber-optic transmission systems via silicon photonics. *Nanophotonics, 9*(16), 4629–4663.
24. Shimobaba, T., Weng, J., Sakurai, T., Okada, N., Nishitsuji, T., Takada, N., & Ito, T. (2012). Computational wave optics library for C++: CWO++ library. *Computer Physics Communications, 183*(5), 1124–1138.
25. Shtaif, M., Antonelli, C., Mecozzi, A., & Chen, X. (2022). Challenges in estimating the information capacity of the fiber-optic channel. *Proceedings of the IEEE, 110*(11), 1655–1678.
26. Smith, F. G., & King, T. A. (2001). Optics and photonics: An introduction. *Measurement Science and Technology, 12*(1), 117–117.
27. Wolf, E. (1954). A macroscopic theory of interference and diffraction of light from finite sources, I. Fields with a narrow spectral range. *Proceedings of the Royal Society of London. Series A. Mathematical and Physical Sciences, 225*(1160), 96–111.
28. Xiao, X., & Voelz, D. (2006). Wave optics simulation approach for partial spatially coherent beams. *Optics Express, 14*(16), 6986–6992.
29. Zhang, X., Li, J., Donegan, J. F., & Bradley, A. L. (2020). Constructive and destructive interference of Kerker-type scattering in an ultrathin silicon Huygens metasurface. *Physical Review Materials, 4*(12), 125202.

CHAPTER

7 Thermodynamics and Statistical Mechanics

LEARNING OBJECTIVES

After studying this chapter, you will be able to:

- Understand the principles of thermodynamic processes, including different types of processes such as isothermal and adiabatic.
- Explain the Laws of thermodynamics (Zeroth, First, Second, and Third Laws) and their implications in understanding energy and entropy.
- Describe the concept of entropy and its relationship with heat in thermodynamic systems.
- Calculate and interpret heat capacity in various systems, including solids, liquids, and gases.
- Apply the Ideal Gas Law to solve problems related to pressure, volume, and temperature relationships in gases.
- Analyze the operation and efficiency of heat engines based on thermodynamic principles.
- Discuss the practical applications of thermodynamics in everyday life and technology.

KEY TERMS FROM THIS CHAPTER

Adiabatic process	Entropy
Heat capacity	Heat engines
Ideal gas law	Isothermal process
Laws of thermodynamics	Thermodynamic processes

UNIT INTRODUCTION

Statistical mechanics originated from the field of thermodynamics, which focused on the thermal characteristics of liquids and gases. The field of classical thermodynamics was formulated and refined over several centuries. However, a significant portion of the foundational research was conducted over a relatively short timeframe, specifically between the 1820s and 1850s.

Statistical mechanics has made significant progress since its modest origins, yet thermodynamics remains a crucial discipline in its own right. This chapter will explore key findings of classical thermodynamics from a contemporary statistical perspective (Sklar, 1999).

7.1. THERMODYNAMIC PROCESSES

Consider a physical system and assume that it is initially in a state of equilibrium. Various modifications can be implemented. Below are a few illustrations.

1. Energy can be introduced into the system directly, for as by agitating it or by exposing it to heat from a fire.
2. It can be exposed to a different thermal reservoir. This can be achieved by either relocating it or by unsealing a door that formerly acted as insulation, so allowing heat to transfer between it and the heat bath.
3. Changing the volume of the system can also add energy. For instance, it could be enclosed within a chamber, with one of its walls connected to a piston that enables it to be displaced inward or outward (Svoboda et al., 2005).
4. The potential energy function for the system can be modified.

Any alteration to the system will disrupt its equilibrium. If sufficient time is allowed for it to rest, nevertheless, it will ultimately return to a state of equilibrium. The range of periods could vary greatly. The range of periods could vary greatly, ranging from milliseconds to several million years. The process of altering the system and enabling it to reach a state of equilibrium is referred to as a thermodynamic process.

So far, two categories of systems have been addressed (Lostaglio et al., 2015):

1. An isolated system with stable and unchanging macroscopic attributes, such as energy and volume.
2. A system in thermal equilibrium with a heat reservoir.

Presently, there exists an additional category: a system with direct control over its properties. They can be transformed according to deliberate choices. This is achieved by implementing external modifications, intervening from an external source to modify the system. Before making the change, the system, which may involve a heat bath, is completely separated from its surroundings. Once the alteration is completed, it is once again separated or kept apart. However, during the process of making the change, it is not isolated (Andresen & Gordon, 1994).

7.2. THE LAWS OF THERMODYNAMICS

Classical thermodynamics is founded upon four laws, which are logically numbered, beginning with zero. When examined through a contemporary lens, three of them are inconsequential. The remaining one is highly significant and is frequently misinterpreted even in contemporary times.

7.2.1. The Zeroth Law of Thermodynamics

If two objects are both in thermal equilibrium with a third object, they are also in thermal equilibrium with each other.

This law was necessary to establish the validity of defining temperature based on measurements obtained from a thermometer. During the initial development of thermodynamics, there was much uncertainty around the precise definition of "temperature." The measurement was based solely on empirical evidence. This law states that when a thermometer reaches a state of equilibrium with two distinct bodies and records the same measurement for each, combining those entities will result in no overall transfer of heat (Shishkov et al., 2018).

7.2.2. The First Law of Thermodynamics

The change in energy of a system during a thermodynamic process is equal to the heat added to the system minus the work done by the system on its surroundings. This can be expressed in differential form:

$$dE = dQ - dW$$

In this equation, (dE) represents the infinitesimal change in energy; (dQ) represents the infinitesimal heat added to the system; and (dW) represents the infinitesimal work done by the system.

This is a simple expression of the principle of energy conservation. However, it is important to note that the pioneers of classical thermodynamics lacked precise knowledge of the nature of "heat." The discovery of its nature as a form of energy and its role in energy conservation necessitated experimental investigation.

Additionally, it is worth noting that the manner in which the statement is formulated reflects the specific issues they aimed to investigate mechanical engines that receive heat and perform work. Dividing all energy transformations into "heat" and "work" is often challenging, particularly in situations involving chemical reactions or the transport of substances between the system and a heat bath. Undoubtedly, energy conservation is a universally applicable principle that holds in all circumstances without exception (Lustig, 2006).

7.2.3. The Second Law of Thermodynamics

In any thermodynamic process, the total entropy can either increase or remain constant, but it can never decrease.

Alternatively, heat transfer will always occur from a body with a lower temperature to a body with a higher temperature.

This law embodies a profound verity concerning the fundamental nature of the cosmos (Lieb and Yngvason, 1999).

7.2.4. The Third Law of Thermodynamics

As the temperature of a system approaches absolute zero, the entropy of the system decreases towards its lowest possible value.

This law was necessary to establish empirical definitions for temperature and entropy. Statistically speaking, it is a factual assertion that the density of states generally increases with energy for almost all physical systems (Abriata & Laughlin, 2004).

7.3. THE SECOND LAW OF THERMODYNAMICS

The Second Law of Thermodynamics is a foundational principle that governs the behavior of entropy in systems isolated from their surroundings. Entropy is a measure of the number of possible configurations that a system can have, commonly expressed as the logarithm of the system's density of states. In the context of isolated systems, this law asserts that entropy tends to increase as time progresses. This implies that systems tend to transition from less probable states to more likely ones (Green and Naghdi, 1977).

This principle seems straightforward when considering commonplace examples. For example, when a container of gas molecules is unsealed, the molecules are more likely to disperse throughout the room rather than remain confined to a single corner of the container. The dispersion of molecules into a larger volume results in a state of higher entropy due to the increased number of possible molecular configurations. In contrast, it is quite unlikely for all molecules to spontaneously returnto their original structure inside the box after dispersing into the room. The disparity in probabilities exemplifies the fundamental concept of the second law: entropy tends to escalate.

Nevertheless, this intuitive understanding of the second law encounters challenges when contemplating the temporal symmetry of physical laws. Physical laws exhibit symmetry with respect to time reversal, as dictated by the principles of CPT invariance (Charge-Parity-Time). By reversing the velocity of all components within a system, the system's trajectory will precisely retrace itself in reverse. This symmetry suggests that there is no intrinsic differentiation between the concepts of "forward" and "backward" in time given a physical standpoint (Lineweaver and Egan, 2008).

One might argue that if entropy consistently increases over time, why is it not possible for it to decrease? Boltzmann's meticulous proof of the second law revolved around this subject. In 1872, he proposed the "H-theorem" which seemed to provide evidence using statistical mechanics. However, this proof assumed that violated the principle of temporal symmetry. This assumption went unnoticed until Josef Loschmidt's argument in 1876, wherein he highlighted that reversing the velocities of all particles might theoretically lead the system back to a state of lower entropy, thereby contradicting the second law (Schneider and Kay, 1994).

Boltzmann revised his stance by proposing that although entropy could occasionally decrease in isolated instances due to statistical fluctuations, many states would demonstrate an overall increase in entropy. This probabilistic interpretation temporarily resolved the paradox. Nevertheless, it prompted further inquiries into the nature of time asymmetry and the statistical foundations of entropy.

Within the realm of contemporary physics, the second law is not viewed as a law that can be definitively proven, but rather as a fundamental concept that is based on statistical mechanics and the probabilistic characteristics of systems. Entropy, a measure of disorder or unpredictability, grows as time progresses because there are significantly more possible arrangements that lead to higher entropy compared to those that result in lower entropy. The observed increase in entropy in real circumstances can be attributed to this statistical inevitability (Sangi & Müller, 2019).

Furthermore, the concept of entropy is closely connected to the arrow of time, which is the perceived direction of the flow of time from the past to the future. Although physical principles exhibit symmetry, the growth in entropy determines the direction that is often referred to as "forward in time." This asymmetry explains phenomena such as irreversibility and the one-way progression of events" for clarity and directness.

Boltzmann's effort to reconcile the directionality of entropy with the symmetrical nature of time highlights a profound philosophical investigation into the fundamental conditions of the cosmos and the role of entropy in the field of cosmology. The Big Bang theory suggests that the universe started with a condition of very low entropy, which helps explain why entropy increases everywhere. From this standpoint, the rise in entropy signifies the enlargement and development of the universe from a highly organized condition to a more chaotic one (Brandao and Plenio, 2008).

7.4. HEAT AND ENTROPY

Consider a thermodynamic process where a small amount of heat dQ is introduced to a system, while ensuring that keeping its volume and other major variables constant, thereby preventing any mechanical work from being done. The first law can be simplified as dE=dQ (Figure 7.1).

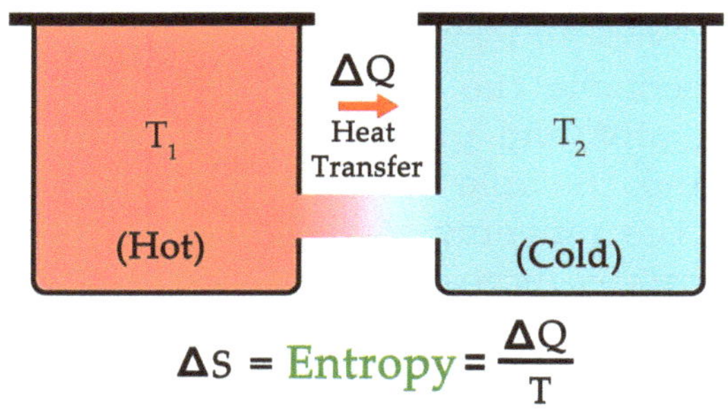

Figure 7.1. *Illustration of heat transfer and entropy.*

Source: Paradigm Cooling, Creative Commons License.

This can be expressed using the chain rule (Saito and Bagchi, 2019):

$$Q = dE = \left(\frac{\partial\, loglog(\Omega)}{\partial E} \right)^{-1} d\left(loglog(\Omega)\right) = TdS \quad (2)$$

In this equation, (Q) represents the heat added to the system; (dE) represents the infinitesimal change in energy; (Ω) is typically the multiplicity or the number of microstates; (T)*is* the temperature and $(S)$is the entropy (Feidt and Costea, 2019).

This demonstrates a clear correlation between heat and entropy. Applying thermal energy to a system increases its entropy. Removing heat from a system decreases its entropy.

Remember

Understanding entropy as a measure of disorder in a system is crucial in thermodynamics, where it dictates the direction of natural processes towards increased randomness.

Assume that there is a transfer of heat, represented by dQ, from subsystem A to subsystem B. The entropy of subsystem A decreases as the entropy of subsystem B increases. The overall entropy change of the entire system is equal to the combined sum of the two individual entropy changes:

$$dS = dS_B + dS_A = \frac{dQ_B}{T_B} - \frac{dQ_A}{T_A}$$

If the temperatures of systems A and B are equal ($T_A = T_B$), then the change in entropy (dS) is zero and the overall entropy of the system remains constant. However, in the scenario where the temperatures are identical, there would be no transfer of heat. When two systems have the same temperatures, they are said to be in thermal equilibrium, meaning that no heat is transferred between them (Singh et al., 2017).

7.5. HEAT CAPACITY

By introducing heat into a system, its energy increases, leading to a corresponding rise in temperature. The heat capacity is the proportionality constant:

$$C \equiv \frac{dQ}{dT} \quad (4)$$

A system with a large heat capacity can absorb a significant amount of heat without causing a substantial change in temperature. Heat capacity is a property that depends on the size of the system. Increasing the size of the system by a factor of two will also increase the heat capacity by a factor of two.

Like the preceding part, assume that the volume remains constant, resulting in no mechanical work being done by the system. Therefore, equation (2) may be used to express the heat capacity at a constant volume (Angell et al., 1982):

$$C_V = \left(\frac{TdS}{dT}\right)_V$$

$$= T\left(\frac{\partial S}{\partial T}\right)_V \quad (5)$$

As the system interacts with the heat bath, its energy and therefore its entropy undergo continual fluctuations.

Therefore, S is not accurately determined in this scenario. However, it is still possible to calculate the average entropy $\langle S \rangle$.

The function of temperature is clearly defined, allowing for the accurate calculation of its derivative with respect to T. The partial derivative used in equation (5) represents the derivative of $\langle S \rangle$, rather than the derivative of S itself.

Initially, entropy was recognized as a measurable quantity by experimentation, but it was subsequently elucidated as a statistical quantity.

From this perspective, there is no ambiguity or confusion associated with any of this. Simply calculate the entropy of a system and see its changes in relation to temperature. Upon observing variations in the outcome based on whether the volume or pressure was controlled in the experiment, appropriately denote the specific measurement by adding a subscript (Sturtevant, 1977).

Did you know?

Heat capacity quantifies the amount of heat needed to change the temperature of a substance. Water has a high heat capacity, making it crucial for moderating temperature changes in ecosystems.

If pressure is manipulated rather than volume, equation (2) will be modified as follows:

$$dE = T\,dS = dQ - P\,dV$$

and the heat capacity becomes

$$\left(\frac{TdS \quad PdV}{dT}\right)$$

$$= \left(T\frac{\partial S}{\partial T}\right)_P + \left(P\frac{\partial V}{\partial T}\right)_P \qquad (6)$$

Empirical evidence demonstrates that in many practical systems, the heat capacity (CP) exceeds the volumetric heat capacity (CV). This is easily comprehensible. When the system expands due to the addition of heat, it exerts energy on its surroundings. This results in a decrease in the energy of the system.

The energy utilized for performing work is not transferred into increasing the temperature, resulting in a slower increase in temperature, denoted as T (Bachmann et al., 1972).

7.6. THE IDEAL GAS LAW

An ideal gas is characterized by a gas composed of non-interacting particles that are free from interaction with one another. It serves as a highly valuable model for examining the behavior of gases during thermodynamic processes.

The ideal gas exhibits a notably simple density of states. The probability of finding a particle in any location within the allotted volume is the same for all particles. Therefore, for an individual particle, the number of possible microstates (Ω) is proportional to the volume (V). The lack of interaction between particles results in the density of states for the entire gas being equal to the product of the density of states for each particle: $\Omega \propto V^N$ (Kautz et al., 2005).

$$P \equiv kT \frac{\partial \, loglog(\Omega)}{\partial V} = NkT \frac{\partial \, loglog(V)}{\partial V} = \frac{NkT}{V}$$

Rearranging this gives the *ideal gas law*:

PV = NkT (7)

An actual gas is inherently distinct from an ideal gas. The particles interact with one another. However, it is noteworthy that this approach proves to be an unexpectedly accurate estimation for dilute gases composed of non-polar, non-reactive molecules. An example would be air. Moreover, even if it lacks sufficient accuracy for making quantitative predictions, it nonetheless provides a reliable qualitative depiction of the behavior of real gases (Li and Fang, 2024).

Consider raising the temperature of a gas, for example, by igniting a fire beneath it. According to the ideal gas law, either the volume, the pressure, or both must increase. When the volume is held constant, the pressure will rise in a linear relationship with the temperature. If the pressure is held constant while permitting the volume to vary, the gas will expand (Lüders and Pohl, 2017).

As it increases in size, it exerts a force on its surroundings according to the equation ⊠=⊠Δ⊠, where Δ⊠ represents the change in volume. Consequently, to maintain energy conservation, the energy of the gas must decrease by an equivalent amount, leading to a subsequent decrease in its temperature. Another significant concept arises from this: when a gas undergoes expansion, its temperature generally decreases (Wang, 2022).

Indeed, it the process functions reciprocally. By exerting work on a gas, one can compress it, which in turn raises its energy level and consequently its temperature.

Thus, it is evident that this uncomplicated gas composed of noninteracting particles possesses a remarkable capability: it can efficiently transfer heat into mechanical work, and vice versa (Laugier & Garai, 2007).

7.7. HEAT ENGINES

A heat engine is a device that converts heat into mechanical work. Based on the observations in the previous section, it appears that a gas could be highly advantageous in constructing one. For instance, consider a scenario where there is a chamber containing gas, and one of the walls of the chamber is a piston that can move back and forth. As the gas undergoes expansion, the piston moves outward and performing mechanical work (Figure 7.2).

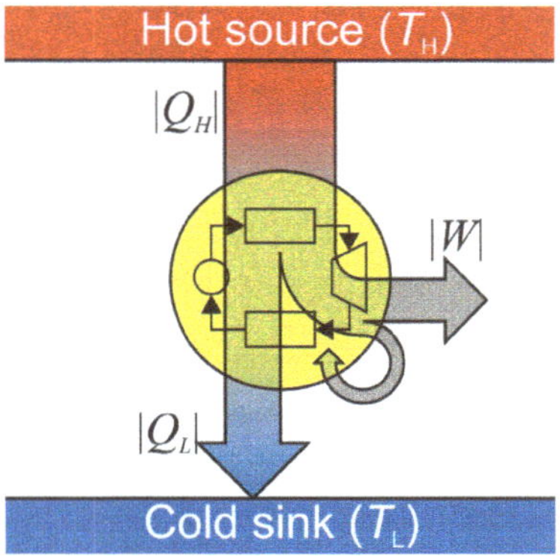

Figure 7.2. *Diagram representing heat engine.*

Source: Gonfer, Creative Commons License.

Once it has reached its maximum extent, it becomes immobile. Before extracting additional work, it is necessary to recompress the gas. To accomplish this, it is necessary to exert effort on the gas, thereby restoring all the valuable work that was previously obtained from it (Gordon & Huleihil, 1992).

When the gas expands, it exerts pressure (P) and causes a change in volume (ΔV) in its surroundings. Conversely, when the gas is compressed, it is needed to exert pressure (P) and cause a change in volume (ΔV) within the gas itself. However, what would happen if the pressures in these two processes were not the same? High pressure is desired during gas expansion and low pressure during compression. Consequently, during the expansion stage, more work can be extracted from the gas compared to the input of work required during the compression stage (Johnson, 2014). The ideal gas law allows this to be achieved by manipulating the temperature. It is desirable for the gas to be heated during expansion and cooled during contraction. This can be readily accomplished. Aside from the gas, which is also referred to as the working body, two distinct heat reservoirs are needed: one known as the "hot bath" or "source" with a temperature of T_H, and another known the "cold bath" or "sink" with a temperature of T_C (Martínez et al., 2017).

1. Place the working object in touch with the hot bath and allow it to undergo expansion. As it expands, it performs valuable work. Typically, this would result in a decrease in its temperature. However, due to being in contact with a heat bath, it consistently absorbs energy and maintains its temperature at T_H.
2. Now remove the working object from the heat reservoir so that it becomes insulated. Allow it to further expand. The isolated system remains productive and experiences a decrease in temperature. Allow this process to persist until the temperature reaches T_C.
3. Place it in touch with the cold bath and initiate compression. Under normal circumstances, this would increase temperature. However, due to its interaction with the cold bath, it releases energy to the bath instead and maintains a temperature of T_C.
4. Detach the functioning component from the bathtub to restore its isolation. Keep working on compressing it. This will now increase its temperature. Repeat until it reaches the temperature threshold, denoted as TH (Kosloff and Levy, 2014).

Tip

Improving the efficiency of heat engines involves minimizing heat loss through insulation and optimizing the thermodynamic cycles used, enhancing energy conversion efficiency.

Examine the events that occur throughout this cycle. Let Q_H represents the heat received from the hot bath, Q_C represents the heat released to the cold bath, and W_W represents the net amount of work produced during the entire cycle. W_W is calculated by subtracting the work done on the working body in steps 3 and 4 from the work done by the working body in steps 1 and 2. The principle of conservation of energy states that the work done on a system is equal to the heat added to the system minus the heat lost by the system.

The entropy of a system varies when there is a transfer of heat inside or outside of the system. Consequently, the entropy of both heat baths must vary throughout the cycle. The entropy of the hot bath decreases while the entropy of the cold bath increases (Holubec & Ryabov, 2021).

The efficiency of a heat engine, denoted by the symbol η, is defined as the quotient of the work perfor med by the engine and the energy absorbed from the hot bath:

$$\eta \equiv \frac{W}{Q_H} = \frac{Q_H - Q_C}{Q_H} = 1 - \frac{Q_C}{Q_H} = 1 - \frac{T_C \Delta S_C}{T_H \Delta S_H} \quad (8)$$

The second rule of thermodynamics mandates that the overall entropy of the entire system must either increase or remain constant during the cycle. The working body ultimately returns to its initial condition, resulting in no change in its entropy. The engine also affects its environment. It is possible that all the energy is concentrated in a single degree of freedom, resulting in zero entropy. Given that only two heat baths are available, the second law of thermodynamics dictates that the change in entropy of the cold bath (ΔS_C) must be greater than or equal to the change in entropy of the hot bath (ΔS_H). Thus (Gutkowicz-Krusin et al., 1978):

$$\eta \leq 1 - \frac{T_C}{T_H} \tag{9}$$

Observe the minimal assumptions made in this derivation: specifically, that the heat engine receives heat from one reservoir, released heat to another reservoir, and performed work. Hence, equation (9) is equally applicable to heat engines employing various cycles (Tu, 2021).

ACTIVITY 7.1.

Conduct an experiment to illustrate the second law of thermodynamics by observing heat transfer from a warmer to a cooler object, demonstrating entropy increase and irreversibility.

SUMMARY

Chapter Seven explores the fundamentals of thermodynamics and statistical mechanics, beginning with thermodynamic processes such as isothermal and adiabatic processes. It covers the laws of thermodynamics, from the Zeroth Law which defines thermal equilibrium to the Third Law that describes absolute zero and entropy. The chapter emphasizes the role of entropy in heat and energy transfer, explaining how it relates to disorder and the direction of spontaneous processes. Heat capacity is discussed in terms of its definition and calculation methods for various substances.

Furthermore, the Ideal Gas Law is introduced and applied to understand the behavior of gases under different conditions of temperature and pressure. Heat engines, their principles of operation, and their efficiency based on thermodynamic cycles are also explored. The chapter concludes with an overview of the practical applications of thermodynamics in heat engines, refrigeration, and other technologies essential to modern life.

REVIEW QUESTIONS

1. Explain the concept of entropy and how it relates to the Second Law of Thermodynamics.
2. Discuss the practical implications of the Third Law of Thermodynamics in achieving absolute zero temperature.
3. Calculate the heat capacity of a substance given its mass, specific heat capacity, and temperature change.
7. Compare and contrast isothermal and adiabatic processes in terms of energy exchange and temperature change.
8. How do heat engines utilize the principles of thermodynamics to convert heat energy into mechanical work?

MULTIPLE CHOICE QUESTIONS

1. The Second Law of Thermodynamics states that:

a. Energy cannot be created or destroyed

b. Entropy of an isolated system always increases

c. All processes proceed spontaneously

d. Heat flows from a cold object to a hot object

2. Entropy is a measure of:

a. The disorder or randomness in a system

b. The total energy of a system

c. The ability of a system to do work

d. The internal pressure of a system

3. Heat capacity is defined as:

a. The amount of heat required to raise the temperature of a substance by one degree Celsius

b. The total heat content of a substance

c. The energy required to convert a substance from solid to liquid state

d. The amount of heat released during a chemical reaction

4. In an isothermal process:

a. The volume of the gas remains constant

b. The pressure of the gas remains constant

c. The temperature of the gas remains constant

d. The internal energy of the gas remains constant

5. A heat engine operates on the principle of:

a. Increasing the entropy of the working substance

b. Decreasing the temperature of the surroundings

c. Converting heat energy into mechanical work

d. Maintaining thermal equilibrium

Answers to Multiple Questions

1. (b); 2. (a); 3. (a); 4. c); 5. (c)

REFERENCES

1. Abriata, J. P., & Laughlin, D. E. (2004). The third law of thermodynamics and low temperature phase stability. *Progress in Materials Science, 49*(3, 4), 367–387.
2. Andresen, B., & Gordon, J. M. (1994). Constant thermodynamic speed for minimizing entropy production in thermodynamic processes and simulated annealing. *Physical Review E, 50*(6), 4346.
3. Angell, C. A., Sichina, W. J., & Oguni, M. (1982). Heat capacity of water at extremes of supercooling and superheating. *The Journal of Physical Chemistry, 86*(6), 998–1002.
4. Bachmann, R., DiSalvo, F., Geballe, T. H., Greene, R. L., Howard, R. E., King, C. N., & Zubeck, R. B. (1972). Heat capacity measurements on small samples at low temperatures. *Review of Scientific Instruments, 43*(2), 205–214.
5. Brandao, F. G., & Plenio, M. B. (2008). Entanglement theory and the second law of thermodynamics. *Nature Physics, 4*(11), 873–877.
6. Feidt, M., & Costea, M. (2019). Progress in Carnot and Chambadal modeling of thermomechanical engine by considering entropy production and heat transfer entropy. *Entropy, 21*(12), 1232.
7. Gordon, J. M., & Huleihil, M. (1992). General performance characteristics of real heat engines. *Journal of Applied Physics, 72*(3), 829–837.

8. Green, A. E., & Naghdi, P. (1977). On thermodynamics and the nature of the second law. *Proceedings of the Royal Society of London. A. Mathematical and Physical Sciences, 357*(1690), 253–270.

9. Gutkowicz-Krusin, D., Procaccia, I., & Ross, J. (1978). On the efficiency of rate processes. Power and efficiency of heat engines. *The Journal of Chemical Physics, 69*(9), 3898–3906.

10. Holubec, V., & Ryabov, A. (2021). Fluctuations in heat engines. *Journal of Physics A: Mathematical and Theoretical, 55*(1), 013001.

11. Johnson, C. V. (2014). Holographic heat engines. *Classical and Quantum Gravity, 31*(20), 205002.

12. Kautz, C. H., Heron, P. R., Loverude, M. E., & McDermott, L. C. (2005). Student understanding of the ideal gas law, Part I: A macroscopic perspective. *American Journal of Physics, 73*(11), 1055–1063.

13. Kosloff, R., & Levy, A. (2014). Quantum heat engines and refrigerators: Continuous devices. *Annual Review of Physical Chemistry, 65*(1), 365–393.

14. Laugier, A., & Garai, J. (2007). Derivation of the ideal gas law. *Journal of Chemical Education, 84*(11), 1832.

15. Li, Z., & Fang, L. (2024). On the ideal gas law for crowds with high pressure. *Physica A: Statistical Mechanics and its Applications, 638*, 129657.

16. Lieb, E. H., & Yngvason, J. (1999). The physics and mathematics of the second law of thermodynamics. *Physics Reports, 310*(1), 1–96.

17. Lineweaver, C. H., & Egan, C. A. (2008). Life, gravity and the second law of thermodynamics. *Physics of Life Reviews, 5*(4), 225–242.

18. Lostaglio, M., Jennings, D., & Rudolph, T. (2015). Description of quantum coherence in thermodynamic processes requires constraints beyond free energy. *Nature Communications, 6*(1), 6383.

19. Lüders, K., & Pohl, R. O. (2017). The First Law and the Equation of State of Ideal Gases. In *Pohl's Introduction to Physics: Volume 1: Mechanics, Acoustics and Thermodynamics* (Vol. 2, pp. 381–408).

20. Lustig, R. H. (2006). Childhood obesity: Behavioral aberration or biochemical drive? Reinterpreting the First Law of Thermodynamics. *Nature Clinical Practice Endocrinology & Metabolism, 2*(8), 447–458.

21. Martínez, I. A., Roldán, É., Dinis, L., & Rica, R. A. (2017). Colloidal heat engines: A review. *Soft Matter, 13*(1), 22–36.

22. Saito, S., & Bagchi, B. (2019). Thermodynamic picture of vitrification of water through complex specific heat and entropy: A journey through "no man's land." *The Journal of Chemical Physics, 150*(5), 89–92.

23. Sangi, R., & Müller, D. (2019). Application of the second law of thermodynamics to control: A review. *Energy, 174*, 938–953.

24. Schneider, E. D., & Kay, J. J. (1994). Life as a manifestation of the second law of thermodynamics. *Mathematical and Computer Modelling, 19*(6–8), 25–48.

25. Shishkov, V. Y., Andrianov, E. S., Pukhov, A. A., Vinogradov, A. P., & Lisyansky, A. A. (2018). Zeroth law of thermodynamics for thermalized open quantum systems having constants of motion. *Physical Review E, 98*(2), 022132.

26. Singh, M., Lara, S. O., & Tlali, S. (2017). Effects of size and shape on the specific heat, melting entropy and enthalpy of nanomaterials. *Journal of Taibah University for Science, 11*(6), 922–929.

27. Sklar, L. (1999). The reduction (?) of thermodynamics to statistical mechanics. *Philosophical Studies: An International Journal for Philosophy in the Analytic Tradition, 95*(1/2), 187–202.

28. Sturtevant, J. M. (1977). Heat capacity and entropy changes in processes involving proteins. *Proceedings of the National Academy of Sciences, 74*(6), 2236–2240.

29. Svoboda, J., Turek, I., & Fischer, F. D. (2005). Application of the thermodynamic extremal principle to modeling of thermodynamic processes in material sciences. *Philosophical Magazine, 85*(31), 3699–3707.

30. Tu, Z. C. (2021). Abstract models for heat engines. *Frontiers of Physics, 16*(3), 33202.

31. Wang, F. (2022). Rethinking the ideal gas law using graphs. *The Physics Teacher, 60*(7), 600–601.

CHAPTER

8

Nuclear Physics

LEARNING OBJECTIVES

After studying this chapter, you will be able to:

- Understand the fundamental terminology and units used in nuclear physics, including physical constants and the concept of nuclear radius.
- Explain the concept of binding energy and its significance in maintaining nuclear stability.
- Describe the components and their roles of the Semi-Empirical Mass Formula used to calculate nuclear binding energies.
- Identify and explain the different types of radioactive decay processes, including alpha and beta decay.
- Analyze the Line of Stability on the Chart of Nuclides and its implications for nuclear stability.
- Discuss the practical applications of nuclear physics in various fields, such as energy production and medical imaging..

KEY TERMS FROM THIS CHAPTER

Alpha decay	Beta decay
Binding energy	Line of stability
Nuclear physics	Nuclear radius
Radioactive decay	Semi-empirical mass formula
Poynting vector	

UNIT INTRODUCTION

This chapter on nuclear physics offers a thorough examination of the fundamental ideas that govern atomic nuclei. It begins with an introduction to important terms, units, dimensions, and physical constants before delving into basic topics such as nuclear radius, binding energy, and radioactive decay. These fundamentals shed light on the forces holding these particles together and the mechanisms driving their transformations, which is essential for comprehending the complex behaviors and interactions within atomic nuclei (Livingston & Bethe, 1937).

8.1. BASICS OF NUCLEAR PHYSICS

This section provides an overview of key notations and fundamental concepts in the field of Nuclear Physics.

8.1.1. Terminology

An atom is uniquely identified by its atomic number.

1. neutrons: N;
2. protons: Z;
3. electrons: Z electrons are present in neutral atoms.

Z is the atomic number shared by all atoms of the same element. But they are not all created equal. Elemental isotopes differ in the number of neutrons (N) (Ahmad, 1996).

Isotopes are commonly represented by the notation (ZAX_N) or simply (ZAX),

Where, The mass number is represented by A, which is equal to the sum of Z and N (A = Z + N). X is the chemical symbol. For example, the isotope of uranium $92235U$ with the atomic number 92 and mass number 235 is written as $92235U$. The atomic number is sometimes omitted because it is unnecessary. Another isotope of uranium with the mass number 238 is written as ^{238}U (Negele & Vogt, 2003).

When discussing various nuclei, they can be referred to as

1. Nuclide: atom or nucleus having a certain Z and N.
2. Isobar: nuclides with identical mass number # A (≠ Z, N)
3. Isotone: identical N nuclides, ≠ Z.
4. Isomer: same nuclide but with a different energy state (Brown, 2005).

8.1.2. Units, Dimensions and Physical Constants

The unit of measurement for nuclear energies is the electronvolt (eV), which is equivalent to $1eV = 1.6 \times 10^{-19} J$. Electronvolts are the unit of measurement for the kinetic energy gained by an electron when it is accelerated by a one-volt potential difference. Nuclear energy often fall within the MeV range, which stands for mega-electronvolt, or 10^6 eV (Pellis, 2022).

The measurement of nuclear masses is expressed in atomic mass units (amu) or unified atomic mass units (u), where 1 amu or $1u = 1.66 \times 10^{-27} kg$. An atomic mass unit (amu) is defined as 1/12th of the mass of a neutral ground-state atom of carbon-12 (12C). Due to the significant difference in weight between electrons and protons/neutrons, with protons and neutrons having similar mass, a single nucleon has a mass of approximately 1 atomic mass unit (amu).

Tip

Understanding nuclear terminology and physical constants helps in interpreting nuclear phenomena and calculations, such as nuclear radius, cross-sections, and decay rates.

Due to the principle of mass-energy equivalence, it is common to represent masses using energy units. The conversion factor between mass (in amu) and energy (in MeV) is the square of the speed of light, as given by the equation $E = mc^2$. Within these units, we have the equation $c^2 = 931.502 MeV / u$ (Bethe, 1937).

2. Proton mass: $938.280 MeV / c^2$
3. Neutron mass: $938.573 MeV / c^2$
4. Electron mass: $0.511 MeV / c^2$

The scales of magnitude for typical lengths are measured in femtometers ($1 fm = 10^{-15} m$), also known as Fermis (F), and Angstroms ($1 A^{\circ} = 10^{-10}$ m) for atomic characteristics. In contrast, typical time scales cover a wide range.

The speed of light, $c = 299,792,458 ms^{-1}$, the electron charge, $e = 1.602176487 \times 10^{-19} C$, The Planck constant, $h = 6.62606896 \times 10^{\wedge} -34 J$ s and $\hbar$, the permittivity of vacuum, $\epsilon_0 = 8.854187817 \times 10^{-}12 Fm^{-1}$ (F=Faraday), Avogadro's number, $N_a = 6.02214179 \times 1023 mol^{-1}$, and some other physical constants will be encountered (Cohen, 1988).

8.1.3. Nuclear Radius

Since a nucleus cannot be described as a rigid sphere with a definite radius, its radius is not well defined. In several experimental settings, such as scattering measurements, it is possible to specify the range at which the density of nucleons converges to a fundamental spherical model. The empirical radius formula is a concise equation that establishes a relationship between the radius of the nucleus and the number of nucleons (Tanihata et al., 1985):

$$R = R_0 A^{1/3}$$

8.2. BINDING ENERGY

It is necessary to study two crucial aspects of nuclear physics: the mass of nuclides and nuclear binding energy. One might assume that a simple formula may be used to compute the masses of all nuclides given the knowledge of the proton and neutron masses:

$$m_N \approx ?\, Z \cdot m_p + N \cdot m_n$$

However, experimental evidence contradicts this assumption. Based on the principles of special relativity theory, it is understood that there is a direct relationship between mass and energy, as expressed by the equation $E = mc^2$. The energy of a nucleus can be calculated by summing the masses of its components. The intrinsic energy and mass of nuclei are linked. We believe that the nucleus's cohesion requires additional energy, therefore this energy is not simply the sum of its component energies If the energy levels were identical, the existence of stable bound nuclei would become unfavorable. Consequently, all nuclei would become inherently unstable, continuously transitioning between their bound state and a combination of protons and neutrons (Bertsch et al., 2005).

Did you know?

The concept of binding energy explains the energy required to disassemble a nucleus into its constituent protons and neutrons, crucial for understanding nuclear stability and reactions.

The difference in mass energy between a nucleus and its components determine the binding energy of a nucleus. In the context of nuclear physics, the binding energy B can be determined for a nucleus $_{ZAX_N}$ using a specific formula.

$$B = \left[Zm_p + Nm_n - m_N(^A X) \right] \cdot c^2$$

Nevertheless, we aim to represent this quantity in terms of measurable and observable values. Nuclear mass is expressed in atomic mass units, a measurable quantity.

$$m_N(^A X)c^2 = \left[m_A(^A X) - Zm_e \right]c^2 + B_e$$

$m_A(^A X)$ represents the atomic mass of the nucleus. To simplify the analysis, we disregard the electronic binding energy B_e by assuming

Below is the equation using similar variables:

$$m_N(^AX)c^2 = \left[m_A(^AX) - Zm_e \right]c^2$$

This equation expresses the rest mass energy of a nucleus (^AX) in relation to its component elements (Figure 8.1).

We have derived the equation for nuclear binding energy (Nikšić et al., 2008).

Here's the equation:

$$B = \{Zm_p + Nm_n - [m_A(^AX) - Zm_e]\} \cdot c^2$$

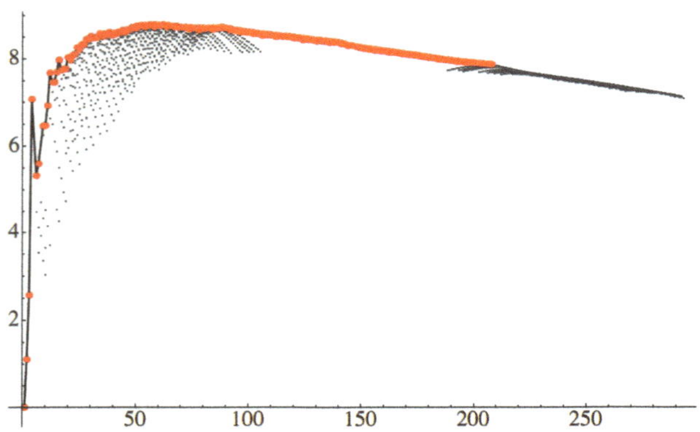

Figure 8.1. *The graph shows the binding energy per nucleon (B/A in MeV vs. A) of both stable nuclides (depicted in red) and unstable nuclides.*

Source: Paola Cappellaro, Creative Commons License.

Another aspect to consider are the separation energies of neutrons and protons, which are of great importance:

$$S_n = B(ZAX_N) - B(ZA-1X_{N-1})$$

$$S_p = B(ZAX_N) - B(Z-1A-1X_N)$$

The corresponding quantities in atomic physics that reflect the energies of valence nucleons are known as ionization energies. It should be noted that these energies reveal information about the nuclei's shell structure (Bender & Heenen, 2011).

8.2.1. Semi-Empirical Mass Formula

The binding energy is commonly represented as B/A or the binding energy per nucleon. Given that the ratio B/A remains largely constant, this indicates that the binding energy is primarily directly related to the atomic mass number.

However, this trend requires some adjustments. The relationship between B/A and A (and Z) is described by the semi-empirical mass formula. This formula is derived from fundamental principles and supported by experimental data to determine the precise parameters that define it (Hein et al., 2022).

The nuclear force evenly distributes and binds the nucleons in the nucleus together in the liquid-drop model. Nevertheless, an existence of the Coulomb contact results in a force of repulsion between the protons. The semi-empirical mass formula can be somewhat explained by considering certain features of the nuclear force, such as its limited range, as well as the Coulomb interaction (Pinedo-Vega et al., 2016).

Additionally, some minor adjustments have been made to consider the fluctuations in binding energy caused by the quantum-mechanical properties, resulting in the development of the nuclear shell model.

The formula for semi-empirical mass (SEMF) is

$$M(Z, A) = Zm(^1H + Nm_n - \frac{B(Z, A)}{c^2}$$

The formula provided gives the binding energy B(Z, A):

$$B(A,Z) = a_v A - a_s A^{2/3} - a_c Z(Z-1) A^{-1/3} - a_{sym}\frac{(A-2Z)^2}{A} + \delta a_p A^{-3/4}$$

volume ↗ surface ↑ Coulomb ↑ symmetry ↑ pairing ↖

Now each term will be studied in the SEMF (Bodmer, 1960).

8.2.2. Volume Term

The initial term represents volume term $a_v A$, which explains relationship between the binding energy and A. What is the reason behind this?

Keep in mind that binding energy measures the interplay between nucleons. Given the dense arrangement of nucleons within the nucleus and the limited reach of the nuclear force, each nucleon ultimately interacts with only a small number of neighboring particles. This indicates that irrespective of the total number of nucleons, each individual nucleon contributes in an equal manner.

Thus, the force is not directly dependent on the number of nucleons that one nucleon can interact with, but rather it is exclusively proportional to A ($A(A-1)/2 \sim A^2$). The constant of proportionality, a_v, is determined through experimental methods and has been found to be 15.5MeV. As the intensity of the nuclear (strong) interaction determines, the value in question is less than the nucleons binding energy to their surrounding particles.

It has been observed (with further study to follow) that the energy required to bind one nucleon to the other nucleons is approximately 50 MeV. The total binding energy is determined by the interaction between nucleons and their neighbors, as well as the kinetic energy of the nucleon itself.

Pauli's exclusion principle states that fermionic nucleons in an atom cannot occupy a similar state with zero kinetic energy. Instead, they will fill all the available energy levels in an atom. In fact, a precise estimate of a_v is provided by this model, which accounts for the kinetic energy resulting from the filling of shells and the nuclear binding energy (Hefter, 1985).

8.2.3. Surface Term

The volume term is adjusted by the surface term, $-a_s A^{2/3}$, which is also dependent on the strong force. We clarified the volume term by stating that, independent of A, every nucleon interacts with a fixed number of other nucleons. This holds true for nucleons located far inside the nucleus, while it is less so for nucleons on the surface of the nucleus This concept is analogous to surface forces, which can be found, for example, in liquid droplets, a process that gives liquids their surface tension.

Since the surface $S \sim V^{2/3}$) and the volume force is proportional to $B_V \propto A$, it is expected that a surface force will be $\sim (B_V)^{2/3}$. In addition, the term should be subtracted from the volume term, and it can be assumed that the coefficient will be of a similar size to that of a_v (Boguta and Bodmer, 1977). Actually, $a_s = 13-18 MeV$.

8.2.4. Coulomb Term

The $-a_c Z(Z-1)A^{-1/3}$ is third term. The Coulomb interaction between protons results in $A^{-1/3}$, which is directly proportional to the atomic number Z. The volume term is reduced by this factor because a nucleus with a high number of protons is less favorable (more energetically unfavorable) due to the Coulomb repulsion. To justify the term's shape and calculate the coefficient a_c, the nucleus is represented as a sphere with uniform charge. A charge distribution of this kind has potential energy that is

$$E = \frac{1}{4\pi\epsilon_0}\frac{3}{5}\frac{Q^2}{R}$$

Considering the unform distribution within the sphere, one can express the charge as $q(r) = \frac{4}{3}\pi r^3 \rho = Q\left(\frac{r}{R}\right)^3$. This allows us to calculate the potential energy.

$$E = \frac{1}{4\pi\epsilon_0}\int dq(\vec{r})\frac{q(\vec{r})}{|\vec{r}|} = \frac{1}{4\pi\epsilon_0}\int d^3\vec{r}\,\rho\frac{q(\vec{r})}{|\vec{r}|} = \frac{1}{4\pi\epsilon_0}\int_0^R dr\,4\pi r^2\rho\frac{q(r)}{r} = \frac{1}{4\pi\epsilon_0}\left(4\pi\int_0^R dr\frac{3Qr^3}{4\pi R^3}Qr^2\left(\frac{r}{R}\right)^3\frac{1}{r}\right) = \frac{1}{4\pi\epsilon_0}\int_0^R dr\frac{3Q^2r^4}{R^6} = \frac{1}{4\pi\epsilon_0}\frac{3}{5}\frac{Q^2}{R}$$

Applying the empirical radius formula $= R_0 A^1/3$ and the total charge $Q^2 = e^2 Z(Z-1)$ (which is relevant when Z > 1, indicating the presence of at least two protons), we obtain the following result:

$$\frac{Q^2}{R} = \frac{e^2 Z(Z-1)}{R_0 A^{1/3}}$$

This equation provides the form of the Coulomb term. The equation can then be used to estimate the constant ac, $a_c \approx \frac{3}{5}\frac{e^2}{4\pi\epsilon_0 R_0}$, with $R_0 = -1.25\,fm$, to be approximately 0.691 MeV, which is quite close to experimental value (Baur and Rebel, 1994) (Figure 8.2).

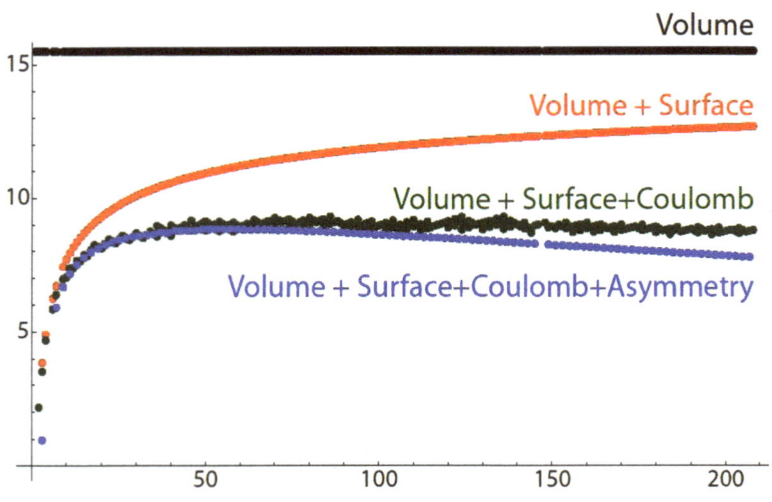

Figure 8.2. *For stable nuclides, use SEMF. We graph A vs. B(Z, A)/A. One by one, the various term contributions are added until the desired final formula is attained.*

Source: Paola Cappellaro, Creative Commons License.

8.2.5. Symmetry Term

According to the Coulomb term, it appears that having a lower number of protons and a higher number of neutrons in a nucleus would be advantageous. Nevertheless, it is evident that the liquid-drop model alone is insufficient to account for the equilibrium of neutrons and protons in stable nuclei. The semi-empirical mass formula (SEMF) contains an additional term that attempts to take the symmetry between neutrons and protons into account (Baldo & Burgio, 2016).

This correction (and the next one) involves the quantum-mechanical exclusion principle and shell model, which will be explored later in the course. Adding additional neutrons would require them to possess greater energy, resulting in an overall increase in the nucleus' total energy. The increase surpasses the coulomb repulsion, making it more favorable to have a nearly equal number of neutrons and protons.

The symmetry term is represented by the formula $\frac{(A-2Z)^2}{A}$, where A represents the atomic mass number and Z represents the number of protons. One can have a clearer understanding by acknowledging that this term approaches zero when A equals 2Z, and its impact becomes less significant for larger A (the symmetry impact becomes more significant for smaller nuclei). The coefficient is $a_{sym} = 23MeV$ (Lattimer & Lim, 2013).

8.2.6. Pairing Term

There is a connection between the last term and the empirical observation that nucleons of the same kind tend to pair up. It can be inferred that in the case of an even-even nucleus, when all protons and neutrons are paired off, the binding energy is greater (δ > 0). One helpful method for examining nuclei with an unequal number of protons and neutrons is to convert one proton into a neutron or vice versa. It is essential to thoroughly analyze the various relevant variables in this context.

Therefore, when all other factors remain constant, it is necessary to subtract a term from the binding energy for configurations with odd-odd nuclei, indicated by a negative value (δ < 0). In the case of even-odd configurations, we can assume that the pairing energy will not have any impact (δ = 0). Next, we can consider the pairing term.

$$\delta_{a_p} A^{-3/4} = \begin{cases} +a_p A^{-3/4} & \text{even-even} \\ 0 & \text{even-odd} \\ -a_p A^{-3/4} & \text{odd-odd} \end{cases}$$

The value of $a_p \approx 34MeV$. [Occasionally, the expression $\propto A^{-1/2}$ is also encountered] (Dean and Hjorth-Jensen, 2003).

8.2.7. Line of Stability in the Chart of nuclides

Through the process of finding the derivative with respect to Z, we can determine the ideal value of Z that minimizes the mass. Here are the results:

$$Z_{min} = \frac{A}{2}\left(\frac{\left(1+\frac{1}{4}A^{-1/3}\frac{a_c}{a_{sym}}\right)}{\left(1+\frac{1}{4}A^{2/3}\frac{a_c}{a_{sym}}\right)}\right) \approx \frac{A}{2}\left(1+\frac{1}{4}A^{2/3}\frac{a_c}{a_{sym}}\right)^{-1}$$

$$\approx \frac{A}{2}\left(1-\frac{1}{4}\frac{A^{2/3}a_c}{a_{sym}}\right)$$

At smaller values of A, the equation yields an approximate value of $Z \approx \frac{A}{2}$. However, for heavier nuclei, there is a correction factor that adjusts the equation to Z ≈ 0.41A. [Please note that the approximation and series expansion are used due to the small value of *ac* compared to *asym*.]

The nuclides fall between 1/2 and 0.41 if we plot D/A. The stable isotopes (black in Figure 8.3 and red in Figure 8.4) are followed by a line of stability (Pons et al., 2013).

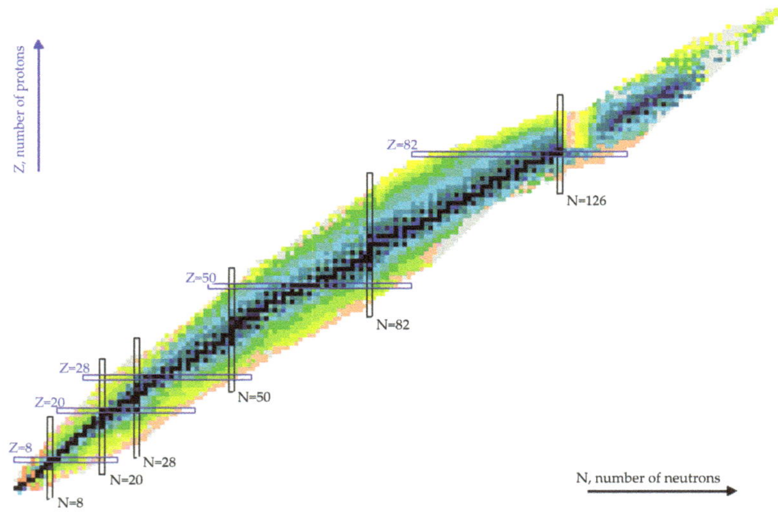

Figure 8.3. *A nuclide chart displaying each nuclide color-coded according to its half-life, with stable nuclides represented in black.*

Source: Brookhaven National Laboratory, Creative Commons License.

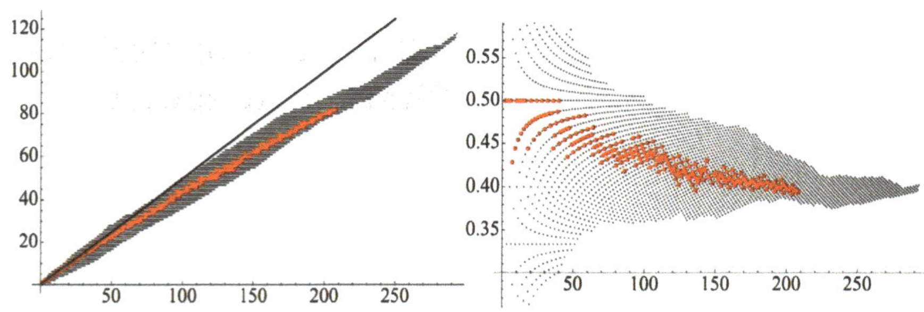

Figure 8.4. *Nuclide chart (produced using the Mathematica program). Right: Z/A against. A; Left: Z vs. A. Stable nuclides in red. Z = A/2 is represented by the black line.*

Source: Paola, Cappellaro, Creative Commons License.

8.3. RADIOACTIVE DECAY

Radioactive decay is the spontaneous loss of energy through the emission of ionizing particles from an unstable nucleus. The process of decay, which involves the dissipation of energy, leads to the conversion of a specific type of atom, known as the parent nuclide, into a different type of atom, called the daughter nuclide.

There are three main modes of decay: beta, alpha, and gamma decays. The variations and mechanisms will be explored later. However, these decay modes have few similar characteristics that will be discussed. The description of these radioactive decays involves quantum processes, specifically transitions between two quantum states.

Therefore, the phenomenon of radioactive decay follows a statistical pattern, which requires us to focus on the expected values of relevant variables, such as the decay rate per unit of time. It is impossible to predict when a single unstable nucleus will decay to a daughter nuclide when monitoring it alone.

The occurrence of decay is random, meaning that at any given moment, there is a probability p that the parent nuclide will decay and a probability of 1 – p that the daughter nuclide will be present. The description of this random event is solely based on the quantum mechanical evolution of the nucleus. Nevertheless, by examining a collection of nuclei, we are able to forecast the average number of both daughter and parent nuclides at any given time.

Let N be the number of radioactive nuclei. The rate of atomic decay per unit time can be represented as the derivative of the number of atoms with respect to time, as .

$$\frac{dN}{dt}$$

It has been observed that this rate is proportional to the number of nuclei present and remains constant throughout time (Atanasov et al., 2015).

$$\frac{dN}{dt} = -\lambda N(t)$$

The value λ, which is known as the decay constant, represents the proportionality factor. Furthermore, the equation mentioned earlier can be restated as

$$\lambda = -\frac{dN}{dt}\frac{1}{N}$$

The right-hand side (RHS) represents the probability of one atom decaying per unit time. The constancy of this probability is a fundamental characteristic of all radioactive decay. Additionally, it results in exponential decay law of radioactive substances:

$$N(t) = N(0)e^{-\lambda}$$

The mean lifetime can also be defined.

$$\tau = \frac{1}{\lambda}$$

Remember

Radioactive decay involves the spontaneous transformation of unstable atomic nuclei into stable nuclei, emitting radiation such as alpha particles, beta particles, or gamma rays.

and the half-life

$$t_{1/2} = \frac{lnln(2)}{\lambda}$$

Which is the duration between the activity and the decay of half of the atoms.

$$A(t) = \lambda N(t)$$

Given that A can also be calculated as $|dN/dt|$, we can estimate the activity by observing the number of decays ΔN during a short time interval δt, where δt is much smaller than the half-life ($t_{1/2}$.).

It is not uncommon for the daughter nuclide to also exhibit radioactive properties. Next, one can observe a series of radioactive decays, each following its own decay laws. For example, in a chain $N_1 \rightarrow N_2 \rightarrow N_3$, the decay of N_1 and N_2 can be described by:

$$dN_1 = -\lambda_1 N_1 dt, dN_2 = +\lambda_1 N_1 dt - \lambda_2 N_2 dt$$

An important characteristic of radioactive decays is their capacity to facilitate the transformation of unstable nuclei into a state that is more energetically advantageous and stable. A nucleus emits an α or β particle during α and β decays to produce a more stable nuclide. On the other hand, γ decay involves the transition of an excited state back to the ground state without altering the nuclear species (Pfützner et al., 2012).

8.3.1. Alpha Decay

An interesting discovery is made while examining the binding energy per mass number (B/A vs. A) plot: there is a noticeable bump or peak in the range of A ~ 60 – 100. These numbers are associated with a minimum energy level. Subsequently, more massive nuclei will naturally tend to undergo decay towards these less massive nuclides, through the release of certain neutrons and protons.

More specifically, Coulomb repulsion is responsible for the binding energy drop at high A. The Coulomb repulsion increases significantly with the square of the atomic number, while the nuclear force only grows in proportion to the atomic mass Z^2 (Royer, 2010).

This can be better understood by drawing a comparison to the fission process, in which a parent nuclide separates into two daughter nuclides. When it comes to α decay, we have the following:

$$ZAX_N \rightarrow Z-2A-4X'_{N-2}+\alpha$$

where α is the nucleus of He-4: $24He_2$

The α decay should be in competition with other processes, such as fission into daughter nuclides of equal mass, or into pairs consisting of ^{12}C and ^{16}O, which have a higher binding energy per nucleon (B/A) than α.

However, α decay is typically preferred. To comprehend this phenomenon, we initiate our analysis by examining the energy aspects of the decay. However, a comprehensive explanation necessitates an investigation into the quantum origins of decay (Koura, 2012) (Figure 8.5).

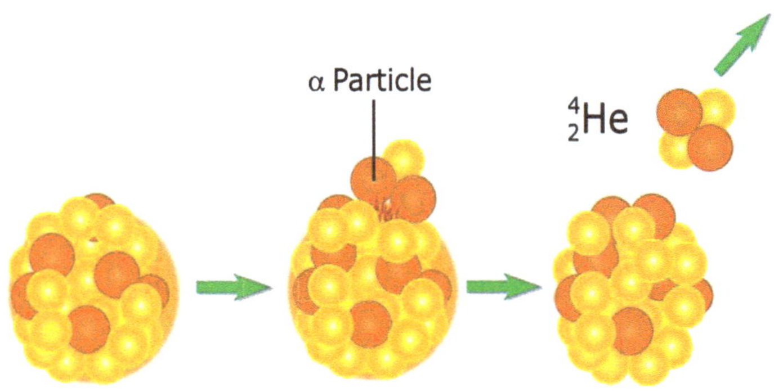

Figure 8.5. *Schematics of alpha decay.*

Source: Ali Abdulwahab Ridha, Creative Commons License.

8.3.2. Beta Decay

One type of radioactive decay is beta decay, which involves the conversion of a proton into a neutron, or vice versa, within a nucleus.

Therefore, the value of A remains constant, although both Z and N increase or decrease by 1. A beta particle, such as a positron or an electron, and a practically massless particle, called a neutrino, are released from the nucleus during beta decay (Šimkovic, 2010) (Figure 8.6).

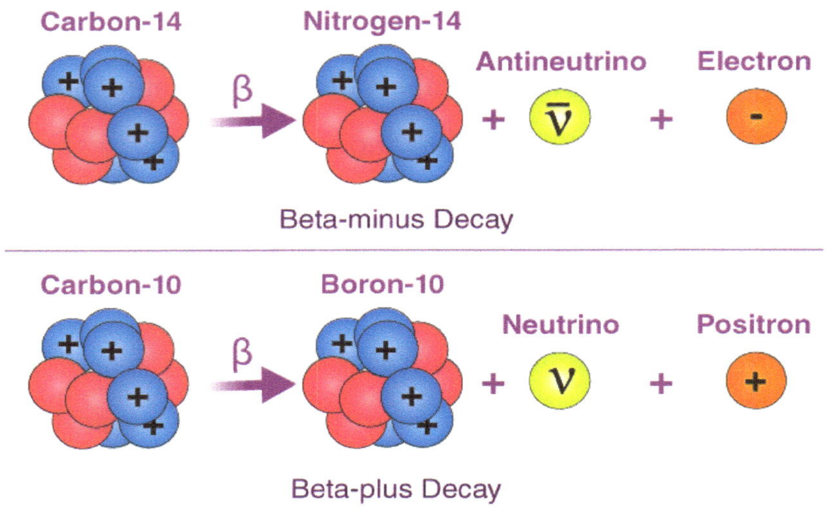

Figure 8.6. *Schematics of beta decay.*

Source: BYJU's, Creative Commons License.

There exist three distinct forms of beta decay:

$$ZAX_N \rightarrow Z+1AX'_{N-1} + e^- + \underline{\nu}$$

This is the process of $\beta-$ decay, often known as negative beta decay. The fundamental reaction is:

$$n \rightarrow p + e^- + \underline{\nu}$$

This refers to the process where a proton is transformed into a neutron by releasing an anti-neutrino and an electron. There are two other sorts of reactions, namely the $\beta+$ reaction,

$$ZAX_N \rightarrow Z-1AX'_{N+1} + e^+ + \nu \Leftrightarrow \ p \rightarrow n + e^+ \ \nu$$

The procedure includes electron capture in addition to the production of a neutrino and positron, which is the anti-particle of an electron:

$$ZAX_N + e^- \rightarrow Z-1AX'_{N+1} + \nu \Leftrightarrow \ p + e^- \rightarrow n \ + \ \nu$$

An alternative method that competes or replaces positron emission (Algora et al., 2021).

Remember that the Nuclide's mass can be given by the semi-empirical mass formula. When A is constant, the SEMF offers the binding energy depending on Z. There is a specific term in the equation that is solely influenced by the value of Z. Upon careful examination, it becomes obvious that $B \propto Z^2$. The Semi-Empirical Mass Formula (SEMF) states that nuclides with the same mass number have a parabolic trend in their masses.

Isotopes located towards the bottom of the parabola exhibit a smaller M value, indicating a higher level of stability. Unstable nuclides change surplus protons into neutrons and vice versa to decay into the minimum required state (Morales, 1999) (Figure 8.7).

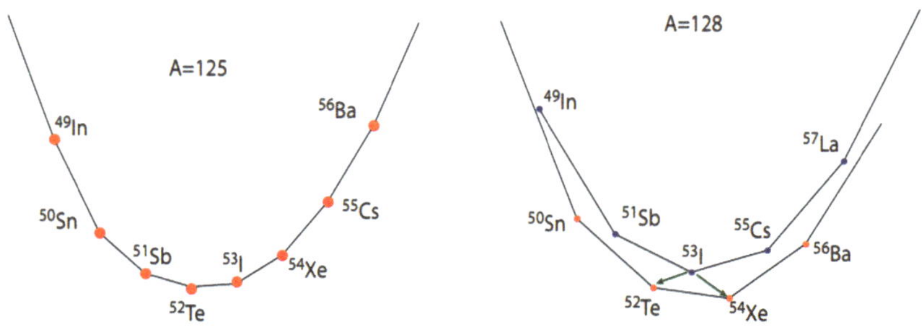

Figure 8.7. *Nuclear Mass Chain for A=128 (right) and A=125, (left).*

Source: MIT Open Course Ware, Creative Commons License.

The process known as beta decay involves the conversion of protons into neutrons and vice versa through radioactive decay. We shall discuss this mechanism later, but for now, let us simply note how it may be energetically beneficial, allowing us to predict transitions based solely on the SEMF.

For example, if A = 125, the stable nuclide with atomic number Z = 52 is tellurium. This is because for atomic numbers less than 52, there is a favorable conversion of neutrons to protons through beta decay. Conversely, for atomic numbers greater than 52, there is a conversion of protons to neutrons by positron emission.

8.3.2.1. Gamma Decay

During gamma decay, the atomic nucleus remains unchanged, but it undergoes a transition from an excited state to a state with reduced energy. These states are commonly known as isomeric states. Typically, the response is expressed as:

$$ZAX_N^* \rightarrow_Z^A X_N + \gamma$$

where an excited state is indicated by a star. The gamma energy is determined by the difference in energy between these two states. However, the decays that can occur are also influenced by the nuclear structure and quantum-mechanical selection criteria for nuclear angular momentum (Vogel, 2012).

8.3.2.2. Spontaneous Fission

Certain nuclei can undergo spontaneous fission even in the absence of the specific circumstances needed in a nuclear reactor. A massive nuclide undergoes nuclear fission, resulting in the formation of two smaller nuclei with approximately equal masses (Baran et al., 2015).

8.3.2.3. Branching Ratios

While some nuclei can only experience one radioactive decay process, others can go through several competing radioactive processes. Branching ratios indicate how strong various decay processes are in relation to one another.

Branching ratios can be expressed either as partial half-lives or as a percentages. For example, if a nucleus can undergo beta decay (along with other decay modes) with a branching ratio of bβ$b\beta$, beta decay partial half-life is given by $\lambda_\beta = b_\beta \lambda$ (Schael et al., 2005).

ACTIVITY 8.1.

Explore the chart of nuclides to identify stable and unstable isotopes, and correlate their positions on the chart with the factors influencing their stability (like neutron-to-proton ratio).

SUMMARY

Chapter Eight introduces the basics of nuclear physics, starting with fundamental terminology, units, and physical constants essential for understanding nuclear phenomena. It covers the concept of binding energy and explores the Semi-Empirical Mass Formula, detailing its components like the volume, surface, Coulomb, symmetry, and pairing terms. The Line of Stability in the Chart of Nuclides is discussed, illustrating the relationship between neutron and proton numbers for stable nuclei.

Furthermore, the chapter delves into radioactive decay processes, including alpha and beta decay mechanisms. It explains how these processes affect nuclear stability and lead to the emission of particles from unstable nuclei. The practical applications of nuclear physics, such as in nuclear reactors for energy production and in medical diagnostics like PET scans, are also highlighted. Overall, this chapter provides a foundational understanding of nuclear physics concepts and their applications in science and technology.

REVIEW QUESTIONS

1. Describe the components of the Semi-Empirical Mass Formula and explain how each contributes to the calculation of nuclear binding energy.
2. What factors determine the stability of nuclei along the Line of Stability in the Chart of Nuclides?
3. Compare and contrast alpha decay and beta decay processes in terms of particle emission and resulting daughter nuclei.
4. Calculate the nuclear binding energy given the mass defect and apply it to determine the stability of a nucleus.
5. Discuss the applications of nuclear physics in medicine and energy production.

MULTIPLE CHOICE QUESTIONS

1. The term used to describe the total energy required to separate a nucleus into its constituent protons and neutrons is:

a. Nuclear Fission

b. Nuclear Fusion

c. Nuclear Binding Energy

d. Nuclear Radius

2. Which term in the Semi-Empirical Mass Formula accounts for the electrostatic repulsion between protons in the nucleus?

a. Volume Term

b. Surface Term

c. Coulomb Term

d. Symmetry Term

3. **Alpha decay involves the emission of:**
   a. Electrons
   b. Protons
   c. Helium nuclei (alpha particles)
   d. Neutrons
4. **The Line of Stability in the Chart of Nuclides represents:**
   a. Nuclei that are stable and do not undergo radioactive decay
   b. Nuclei that are unstable and undergo alpha decay
   c. Nuclei that are unstable and undergo beta decay
   d. Nuclei that are in equilibrium with their environment
5. **Nuclear reactors primarily use the process of _____ to generate energy.**
   a. Alpha decay
   b. Beta decay
   c. Nuclear fission
   d. Nuclear fusion

Answers to Multiple Questions

1. (c); 2. (c); 3. (c); 4. (a); 5. (c)

REFERENCES

1. Ahmad, K. (1996). A terminology dynamic and the growth of knowledge: A case study in nuclear physics and in the philosophy of science. In *TKE* (Vol. 96, pp. 1–11).
2. Algora, A., Tain, J. L., Rubio, B., Fallot, M., & Gelletly, W. (2021). Beta-decay studies for applied and basic nuclear physics. *The European Physical Journal A, 57*(3), 85.
3. Atanasov, D., Blaum, K., Bosch, F., Brandau, C., Buehler, P., Chen, X., & Zhang, Y. (2015). Between atomic and nuclear physics: Radioactive decays of highly-charged ions. *Journal of Physics B: Atomic, Molecular and Optical Physics, 48*(14), 144024.
4. Baldo, M., & Burgio, G. F. (2016). The nuclear symmetry energy. *Progress in Particle and Nuclear Physics, 91*, 203–258.
5. Baran, A., Kowal, M., Reinhard, P. G., Robledo, L. M., Staszczak, A., & Warda, M. (2015). Fission barriers and probabilities of spontaneous fission for elements with $Z \geq 100$. *Nuclear Physics A, 944*, 442–470.
6. Baur, G., & Rebel, H. (1994). Coulomb dissociation studies as a tool of nuclear astrophysics. *Journal of Physics G: Nuclear and Particle Physics, 20*(1), 1–5.
7. Bender, M., & Heenen, P. H. (2011). What can be learned from binding energy differences about nuclear structure: The example of δV pn. *Physical Review C—Nuclear Physics, 83*(6), 064319.
8. Bertsch, G. F., Sabbey, B., & Uusnäkki, M. (2005). Fitting theories of nuclear binding

energies. *Physical Review C—Nuclear Physics, 71*(5), 054311.

9. Bethe, H. A. (1937). Nuclear physics B. Nuclear dynamics, theoretical. *Reviews of Modern Physics, 9*(2), 69.
10. Bodmer, A. R. (1960). Neutron and proton densities in nuclei and the semi-empirical mass formula. *Nuclear Physics, 17*, 388–420.
11. Boguta, J., & Bodmer, A. R. (1977). Relativistic calculation of nuclear matter and the nuclear surface. *Nuclear Physics A, 292*(3), 413–428.
12. Brown, B. A. (2005). Lecture notes in nuclear structure physics. *National Super Conducting Cyclotron Laboratory, 11*, 985–990.
13. Cohen, E. R. (1988). Fundamental physical constants. In *Gravitational Measurements, Fundamental Metrology and Constants* (Vol. 2, pp. 59–89). Dordrecht: Springer Netherlands.
14. Dean, D. J., & Hjorth-Jensen, M. (2003). Pairing in nuclear systems: From neutron stars to finite nuclei. *Reviews of Modern Physics, 75*(2), 607.
15. Hefter, E. F. (1985). Application of the nonlinear Schrödinger equation with a logarithmic inhomogeneous term to nuclear physics. *Physical Review A, 32*(2), 1201.
16. Hein, M., Pusch, A., & Heusler, S. (2022). Modeling in nuclear physics: A visual approach to the limitations of the semi-empirical mass formula. *European Journal of Physics, 43*(3), 035801.
17. Koura, H. (2012). Phenomenological formula for alpha-decay half-lives. *Journal of Nuclear Science and Technology, 49*(8), 816–823.
18. Lattimer, J. M., & Lim, Y. (2013). Constraining the symmetry parameters of the nuclear interaction. *The Astrophysical Journal, 771*(1), 51.
19. Livingston, M. S., & Bethe, H. A. (1937). Nuclear physics C. Nuclear dynamics, experimental. *Reviews of Modern Physics, 9*(3), 245.
20. Morales, A. (1999). Review on double beta decay experiments and comparison with theory. *Nuclear Physics-Section B-PS-Proceedings Supplements, 77*, 335–345.
21. Negele, J. W., & Vogt, E. (2003). *Advances in Nuclear Physics* (Vol. 27, pp. 21–25). Springer Science & Business Media.
22. Nikšić, T., Vretenar, D., & Ring, P. (2008). Relativistic nuclear energy density functionals: Adjusting parameters to binding energies. *Physical Review C—Nuclear Physics, 78*(3), 034318.
23. Pellis, S. (2022). Unity formulas for the coupling constants and the dimensionless physical constants. *Journal of High Energy Physics, Gravitation and Cosmology, 9*(1), 245–294.
24. Pfützner, M., Karny, M., Grigorenko, L. V., & Riisager, K. (2012). Radioactive decays at limits of nuclear stability. *Reviews of Modern Physics, 84*(2), 567–619.
25. Pinedo-Vega, J. L., Ríos-Martínez, C., Talamantes-Carlos, M. P., Mireles-García, F., Dávila-Rangel, J. I., & Badillo-Almaraz, V. (2016). Semi-empirical nuclear mass formula: Simultaneous determination of 4 coefficients. *Asian Journal of Physical and Chemical Sciences, 1*(2), 1–10.

26. Pons, D. J., Pons, A. D., & Pons, A. J. (2013). Explanation of the Table of Nuclides: Qualitative nuclear mechanics from a NLHV design. *Applied Physics Research, 5*(6), 145–174.
27. Royer, G. (2010). Analytic expressions for alpha-decay half-lives and potential barriers. *Nuclear Physics A, 848*(3, 4), 279–291.
28. Schael, S., Barate, R., Bruneliere, R., De Bonis, I., Decamp, D., Goy, C., & Foster, F. (2005). Branching ratios and spectral functions of ⊠ decays: Final ALEPH measurements and physics implications. *Physics Reports, 421*(5, 6), 191–284.
29. Šimkovic, F. (2010). Double beta decay: A problem of particle, nuclear and atomic physics. *Progress in Particle and Nuclear Physics, 64*(2), 219–227.
30. Tanihata, I., Hamagaki, H., Hashimoto, O., Shida, Y., Yoshikawa, N., Sugimoto, K., & Takahashi, N. (1985). Measurements of interaction cross sections and nuclear radii in the light p-shell region. *Physical Review Letters, 55*(24), 2676.
31. Vogel, P. (2012). Nuclear structure and double beta decay. *Journal of Physics G: Nuclear and Particle Physics, 39*(12), 124002.

Index

V

W

Y

Z

FSC
www.fsc.org
MIX
Papier | Fördert
gute Waldnutzung
FSC® C083411

Kolibri 360 GmbH
Ferdinand-Jühlke-Straße 7
99095 Erfurt, Deutschland
produktsicherheit@kolibri360.de